Altamaha

Altamaha *A River and Its Keeper*

PHOTOGRAPHS BY James Holland TEXT BY DORINDA G. DALLMEYER AND JANISSE RAY

THE UNIVERSITY OF GEORGIA PRESS *Athens & London*

Athens, Georgia 30602
www.ugapress.org

Maps by Deborah Reade Design
Designed by Erin Kirk New
Set in Minion Pro and Whitney
Printed and bound by Four Colour Print Group
The paper in this book meets the guidelines for permanence and durability of the Committee on Production Guidelines for Book Longevity of the Council on Library Resources.

Printed in China
16 15 14 13 12 P 5 4 3 2 1

Library of Congress Cataloging-in-Publication Data

Holland, James, 1940-
Altamaha : a river and its keeper / photographs by James Holland ; text by Dorinda G. Dallmeyer and Janisse Ray.
p. cm. — (A Wormsloe Foundation nature book)
ISBN-13: 978-0-8203-4312-9 (pbk. : alk. paper)
ISBN-10: 0-8203-4312-9 (pbk. : alk. paper)
1. Natural history—Georgia—Altamaha River—Pictorial works. 2. Altamaha River (Ga.)—Pictorial works. 3. Natural history—Georgia—Altamaha River. 4. Natural history—Georgia—Altamaha River Region. 5. Altamaha River (Ga.)—History. 6. Altamaha River (Ga.)—Description and travel. 7. Altamaha River (Ga.)—Environmental conditions. 8. Holland, James, 1940- 9. Environmentalists—Georgia—Biography. 10. Photographers—Georgia—Biography. I. Title.
QH105.G4H65 2012
508.758'7—dc23 2011044237

British Library Cataloging-in-Publication Data available

To the youth of Georgia.

Here, children, here. Here it is.

Contents

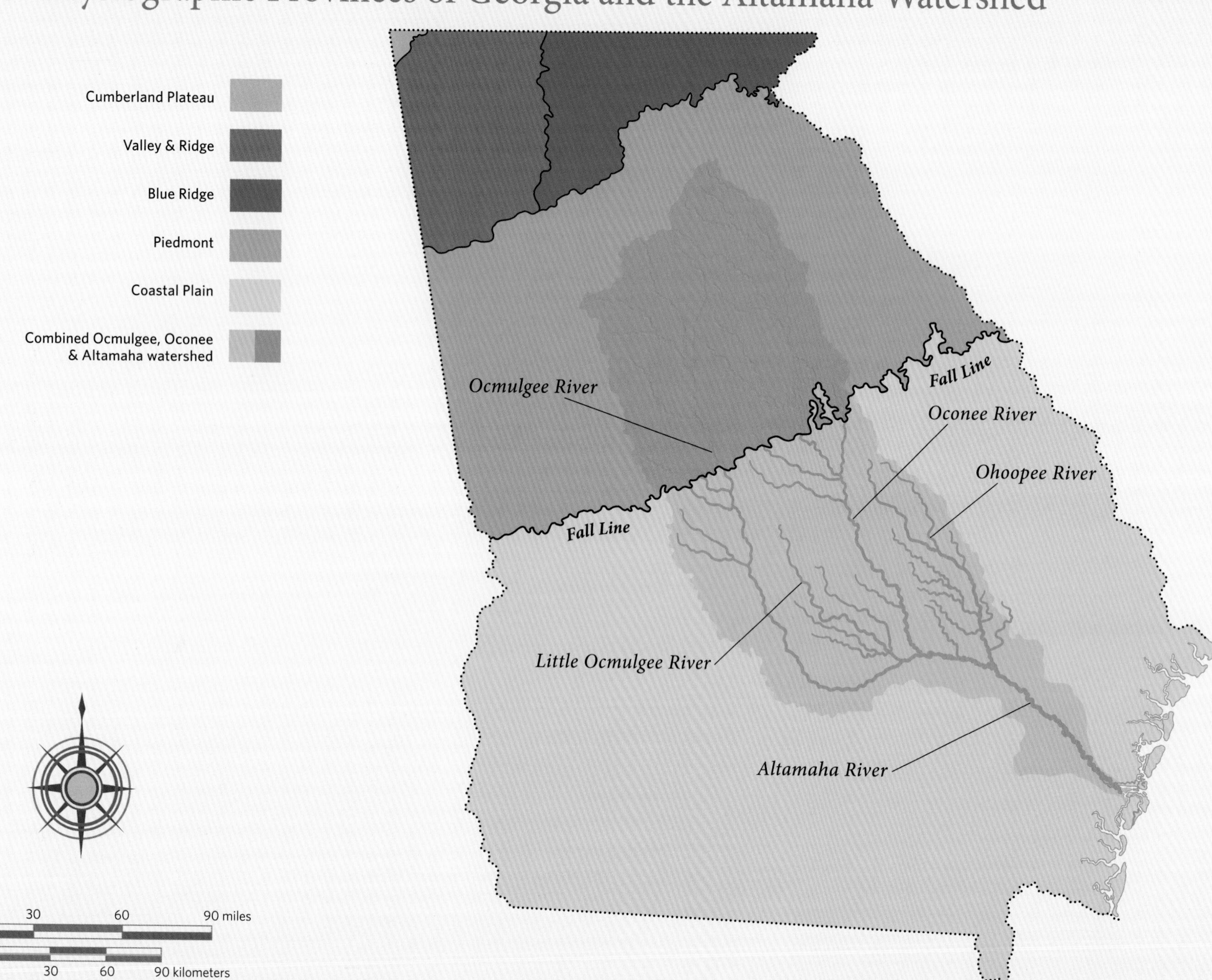
Physiographic Provinces of Georgia and the Altamaha Watershed
Cumberland Plateau
Valley & Ridge
Blue Ridge
Piedmont
Coastal Plain
Combined Ocmulgee, Oconee & Altamaha watershed
Ocmulgee River
Fall Line
Oconee River
Ohoopee River
Fall Line
Little Ocmulgee River
Altamaha River
0
30
60
90 miles
0
30
60
90 kilometers

The Altamaha River and Its Principal Tributaries

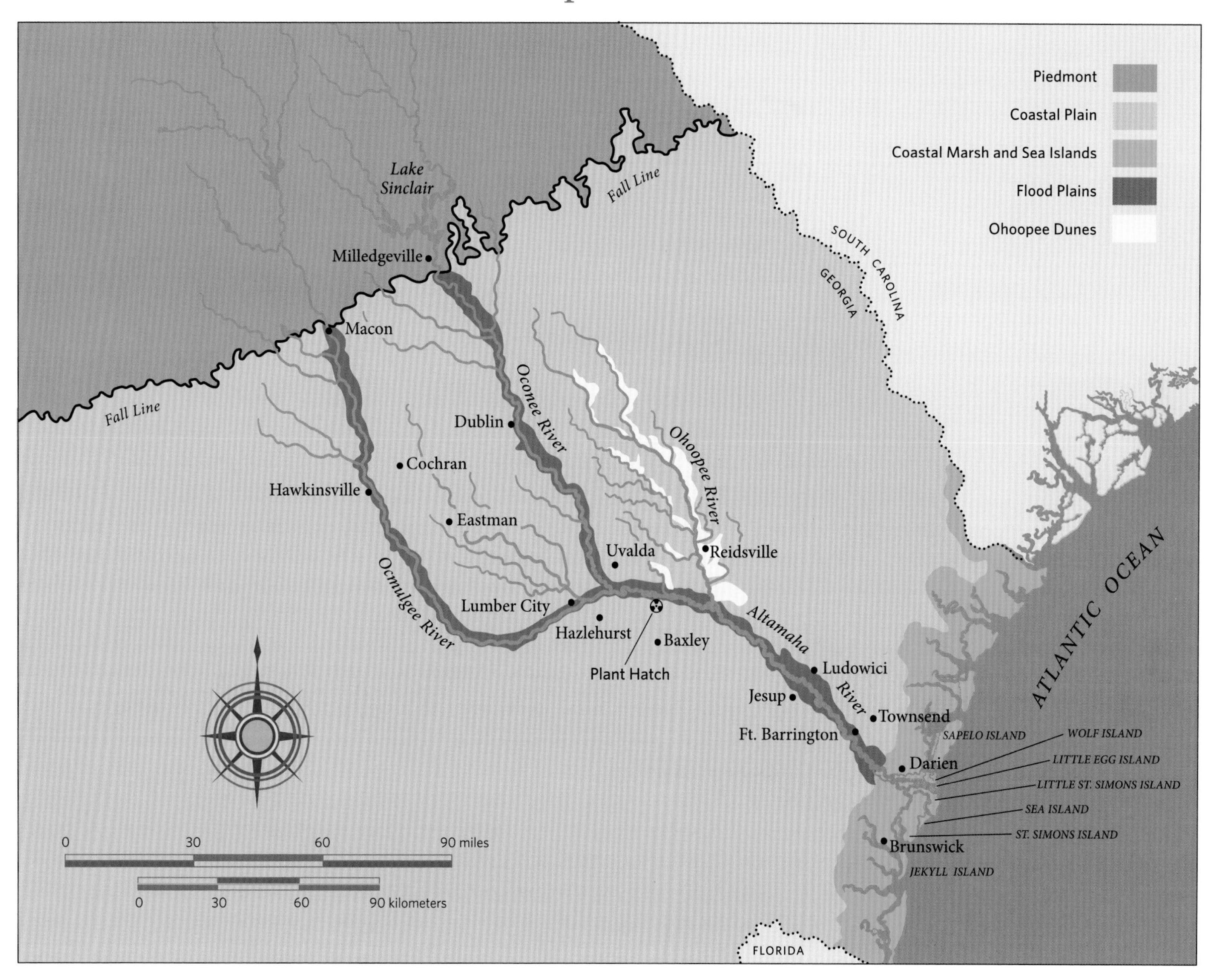

Altamaha *A River for All*

DORINDA G. DALLMEYER

Unlike most of Georgia's warblers, which migrate to Central and South America for the winter, the pine warbler (*Dendroica pinus*) is a year-round resident. Birds migrating from the more northerly parts of its range swell the winter population of the Altamaha, where they feast on the abundant pine seeds. Often seen visiting feeders or gleaning insects in the pines that give it its name, this pine warbler warms in the October sun within a spray of bushy aster (*Aster dumosus*).

As long as there has been this land we call Georgia, rivers have washed its face. Rocks exposed in the Georgia Blue Ridge near Lake Allatoona formed from river sediments more than a billion years ago. Approximately 325 million years ago, as Georgia's Appalachians began to be pushed up to reach the height of today's Rocky Mountains, tumultuous mountain streams immediately began to erode their channels upstream and down. Over millions of years, these unnamed rivers reduced the Appalachians to their familiar rounded profile and in places wore the mountains down to their roots, what we know today as the Piedmont. During the last 100 million years, as global climate change caused the sea to rise and fall across the land, rivers deposited vast loads of sand, silt, and clay at the shore to help form Georgia's Coastal Plain, that great wedge of sedimentary rocks blanketing the southern third of the state.

Whatever these ancestral rivers accomplished in the past, the Altamaha is their worthy descendant. Variously rendered over nearly five centuries as "Altama," "Alatamaha," and even "Hallatomahaw," the name was first mentioned in chronicles describing the De Soto expedition in 1540, where *Al Tama* refers not to the river but to the broader region surrounding the Indian village of Tama southeast of today's Milledgeville. Formed by the confluence of the Oconee and Ocmulgee Rivers, the Altamaha drains the third largest watershed on the eastern seaboard. With the Oconee's headwaters stretching into the foothills of the Blue Ridge and the Ocmulgee's into the heart of south Atlanta, the Altamaha drainage basin covers approximately 14,000 square miles, representing more than one-quarter of the state. Because its 137-mile long course lies wholly within the gently sloping Coastal Plain, the river is a poor candidate for dams. As a result, the Altamaha remains the longest free-flowing river system on the East Coast.

Given its extensive drainage basin, it's not surprising that the Altamaha also has the largest river discharge south of the Chesapeake Bay. On average, the Altamaha delivers over nine billion gallons of fresh water each day to the ocean beyond its mouth between Sapelo Island and Little St. Simons Island. If that volume of water is hard to visualize, imagine ten Olympic-size swimming pools passing by each minute or the volume of thirty-three Empire State Buildings every twenty-four hours.

But the Altamaha is far more than the water it carries. From the confluence to the sea, its valley features a great diversity of habitats: forested swamps and bottomlands, shaded bluffs, isolated sand ridges, and freshwater wetlands that are succeeded in turn by tidal marshes and barrier islands. At periods of high discharge, the river's flow even affects currents circulating on the continental shelf at least as far as the Gray's Reef National Marine Sanctuary sixteen miles offshore. The Altamaha system presents such a unique complex of habitats that the Nature Conservancy named it one of seventy-five "Last Great Places on Earth."

A broad river lined with stands of enormous hardwoods, magnolias, and cypress trees may be a visitor's first impression of the Altamaha, but there is so much more. The river basin supports 234 species of rare plants and animals, some occurring nowhere else on earth, and hosts the highest documented number of ecological communities in Georgia. This habitat diversity attracts at least 160 species of birds that use the Altamaha as summer breeding grounds, wintering grounds, or as a stopover point to rest and refuel during migration. The native plant population is diverse as well. For example, several species of trillium found in the Altamaha basin more typically occur much farther north in the Blue Ridge Mountains. Botanists believe these plants expanded their range southward during past glacial periods. With the return to a warmer climate, the Altamaha plants are remnants of the ice-age populations, able to survive only on cool, shady, north-facing bluffs.

Much less visible are rare aquatic animals found beneath the Altamaha's dark waters. The river hosts the largest population of shortnose sturgeon and Atlantic sturgeon in the Southeast, and perhaps the largest population of Atlantic sturgeon anywhere in the United States. In addition to rare fish, the river is home to seven of Georgia's eight endemic mussel species, species found nowhere else in the world. Many of these members of the bivalve family are in decline not only because of water pollution and sedimentation but also because they depend on specific fish to harbor larval mussels during an important stage of their development. When the host fish disappear, the mussels disappear as well. One of these endemic mussels is the striking *Elliptio spinosa*, commonly called the Altamaha spinymussel because of inch-long spines protruding from each valve of its shell. At one time, the mussel was so abundant along the Altamaha that people wading the river complained about the unpleasant surprise of being "stung" time and again by mussels living on submerged sandbars. Now wildlife biologists are in a race against the clock to understand the reproductive biology of the Altamaha's endangered mussels in hopes of preventing them from vanishing from the river for all time.

The Altamaha offers even more than rich biodiversity and aesthetic beauty. Few people understand that the river also provides life-support services fundamental for human beings. The Altamaha purifies water through decomposition of wastes by chemical as well as biological action. It recycles and moves nutrients, regenerates soil fertility, and produces a wide array of wild game, timber, and seafood. Its forested bottomlands, wetlands, and marshes slow the flow of floodwaters, thereby protecting property, recharging groundwater, and trapping sediment. Its coastal marshes and barrier islands blunt the impact of storms and hurricanes. Its forests help moderate temperature extremes, recycle water within the system, and also sequester carbon. While our current economic system poorly accounts for the values and services that ecosystems provide both in the short and long term, the Altamaha does all this at a scale that would be difficult to replicate even with modern technology, and it does it all for free.

In addition to supporting the needs of a rich variety of living organisms, the Altamaha passes through a remarkable diversity of habitats, many created by the river itself. As the river meanders toward the sea across the gently sloping Coastal Plain, it drops only ninety vertical feet in 137 miles. Along its path, forested swamps and bottomlands provide important habitat for terrestrial wildlife. But the river occasionally floods its banks, and when that happens, fish and other

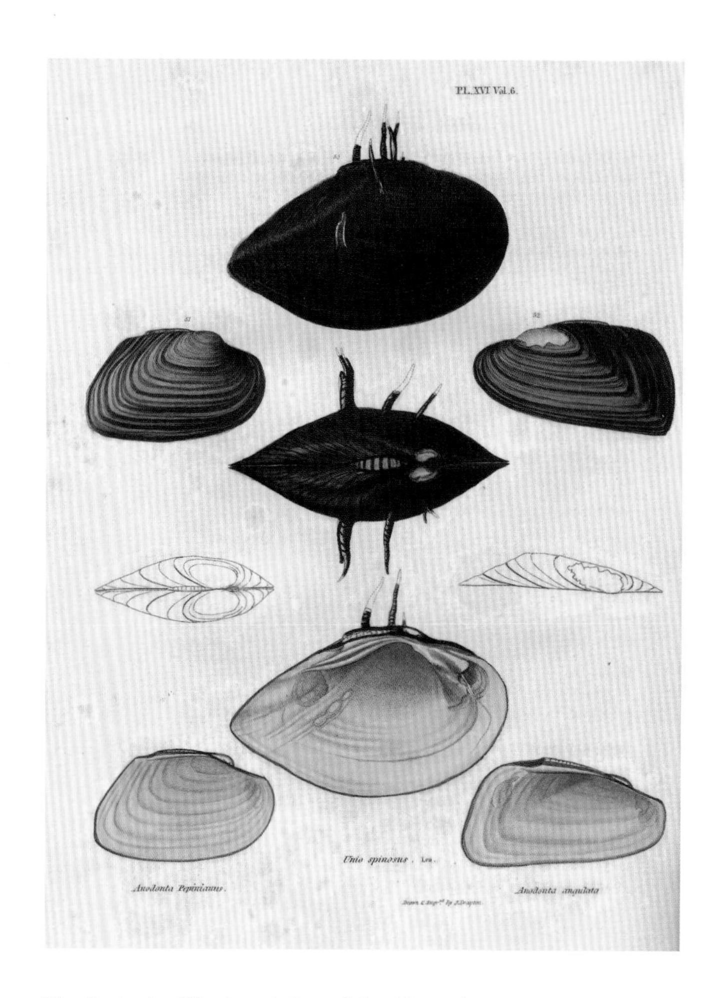

The first scientific description of the Altamaha spinymussel, *Elliptio spinosa,* appeared in the *Transactions of the American Philosophical Society* in 1839, where it was initially named *Unio spinosus.*

aquatic creatures find ideal spawning habitat and have a competitive advantage over predators. Flooded bottomlands offer juvenile fish as well as other vertebrate and invertebrate young the opportunity to exploit rich food sources and to grow rapidly. Once they ride the retreating floodwaters back to the Altamaha, they are better able to compete in their river home.

Just as floods come and go, the very course of the Altamaha changes. Erosion and transport of sediment downstream causes the river to meander so extravagantly that the river eventually cuts through the base of each loop, thereby creating an oxbow lake. Isolated from the rest of the river except during floods, these lakes are famous in the angling community for record largemouth bass and other game fish. They also provide important habitat for nongame fish, waterfowl, aquatic mammals such as otter and beaver, as well as a wide variety of reptiles and invertebrates.

Farther down the river, small tributaries originating wholly within the Coastal Plain augment the Altamaha's flow. These small, slow-moving rivers and creeks are referred to as "blackwater" streams because their waters are stained brown with tannic acids released from decaying vegetation in the heavily forested areas they drain. In addition to surface runoff, these streams also receive a portion of their flow from groundwater sources. The Ohoopee River, the largest tributary on the lower Altamaha, features a number of fish species that prefer blackwater streams, including the eastern silvery minnow, American eel, mud sunfish, sailfin shiner, and sawcheek darter. An unusual feature lies along the Ohoopee's eastern shore: twenty-two thousand acres of sand hills deposited by strong westerly winds that blew across exposed river sands at the height of the last ice age approximately twenty thousand years ago. These fast-draining, nutrient-poor soils of the Ohoopee dunes stunt the growth of trees and shrubs. But rare plants that can cope in this humid desert abound.

Other sand deposits cutting across the Altamaha's path through the lower Coastal Plain are the work of the ocean, not the wind. Just as we have barrier islands off the Georgia coast today, so it was in the past. Although barrier islands may seem like solid ground, functionally they are nothing more than piles of sand deposited offshore, free to roll inland when sea level rises as the climate warms and ice caps melt. In this case, however, what goes up cannot come back down. When the climate swings back into a glacial period, with water again stored as ice in glaciers and ice caps, sea level falls, leaving behind a chain of accreted barrier island sands. In Georgia, each line of stranded islands forms a marine terrace, each recording successive high stands of sea level over the last 120,000 years; the oldest terrace lies nearly one hundred feet above current sea level. The Altamaha cuts through these paleo-shorelines as they stairstep toward the coast and its swamps isolate the sand ridges, protecting them from fire, the other major natural force helping to shape the Altamaha ecosystem. As a consequence, forests flourish on the ridges and wading birds such as wood storks, herons, and egrets use the wetlands for nesting and roosting sites.

Perhaps the sand ridge community's most famous resident is the shrub *Franklinia alatamaha*, named to honor both Benjamin Franklin and the river system that supported the only known population ever found. Pioneer naturalist William Bartram described the showy flowering shrub he saw in the summer of 1776, near Fort Barrington in McIntosh County. He wrote:

> It is a flowering tree, of the first order for beauty and fragrance of blossom; . . .the flowers are very large, expand themselves perfectly, are of a snow-white colour, and ornamented with a crown or tassel of gold-colored refulgent staminae in their centre. . . .We never saw it grow in any other place, nor have I seen it growing wild, in all my travels from Pennsylvania to . . . the banks of the Mississippi.

Fortunately he sent seeds back to his family's botanical garden near Philadelphia where the Bartrams succeeded in cultivating *Franklinia* and sent specimens to plant collectors around the world. Despite diligent searches, botanists have not seen *Franklinia* in its wild Altamaha home since 1803. According to Bartram's description, there were not very many specimens to begin with; the shrub may have been over-collected following his discovery, or it could have been eliminated by some natural event. So every *Franklinia* we see today exists because of Bartram's small gesture of saving its seeds.

Along the lower reaches of the Altamaha floodplain, the tide begins to make its presence felt, but not by making the water salty; the ocean is too far away and the volume of water the Altamaha discharges is too large. Instead, because freshwater is less dense than salt water, the Altamaha's water rides on top of a wedge of salt water, rising and falling as the salt water moves upstream and downstream with each tidal cycle. Here the river forest is dominated by bald cypress and tupelo trees, which are better adapted than other kinds of trees to withstand long periods of inundation and saturated soils. Where loggers failed to clear-cut these stands, giant specimens of old-growth cypress and tupelo abound. In 2010 in a swamp near Townsend, Georgia, Altamaha Riverkeeper James Holland located Georgia's

William Bartram's watercolor of *Franklinia alatamaha* painted in 1788.
© Natural History Museum, London, England.

largest recorded bald cypress, officially measured at forty-four feet, five inches in circumference. Although the age of this tree has yet to be determined, bald cypress are known to live more than one thousand years. Around the looming cypress, colossal cypress knees twice the height of a person combine with the weirdly contorted trunks and branches of ancient Ogeechee tupelo to enhance the fantasy of this landscape.

Even cypress and tupelo cannot tolerate constant flooding, and thus these forests give way downstream to tidally-influenced freshwater marshes dominated by grasses, sedges, rushes, cattails, and even native wild rice. Here as the waters of the Altamaha begin to meet the ocean, the river widens and slows its pace to form an estuary protected from the open sea by a chain of barrier islands lying offshore. The estuary serves as a large filter for the silt and clay brought from far inland, which settles out to form marsh mud. This rich sediment supports the growth of a high diversity of plant species, but the soil's fertility also makes it attractive for agriculture. Consequently, pristine freshwater marsh is one of the rarest habitats remaining on the Georgia coast.

If we drift along the lower Altamaha today, we see the relict floodgates and dikes of rice plantations that flourished in Georgia's tidewater for nearly a century before the Civil War. Following the legalization of slavery in the Georgia colony in 1751, planters used massive amounts of slave labor to convert thirty thousand acres of freshwater marsh into diked rice fields. In a narrow zone ten to twenty miles wide, these plantations combined the power of the natural rise and fall of the tides with a carefully engineered system of dikes and floodgates to suit the varying water demands of rice throughout its growth cycle. By the 1860s, plantations in the tidewater region of Georgia and South Carolina supplied nearly ninety percent of the rice produced in the United States. But rice culture dramatically declined in the wake of the Civil War due to a combination of emancipation, low-country hurricanes, imports of less-expensive rice from Asia, and the westward shift of American rice production. Although very little of the original freshwater marsh remains, the old Altamaha rice fields that at one time spread across islands near Darien now help support thousands of waterfowl as well as other wetland-dependent species.

Beyond the freshwater marshes lies one of the Altamaha's most beloved gifts: the coastal salt marsh. Although this type of salt marsh ranges from North Carolina to northern Florida, the Georgia coast accounts for nearly thirty-six percent of the total found in the Southeast. Georgia also experiences some of the highest tidal ranges south of Cape Cod—up to a twelve-foot rise and fall twice a day during full and new moons (i.e., spring tides). Because of salinity contrasts, tidal flooding controls the distribution of plant species within the salt marsh. *Spartina alterniflora*, known commonly as smooth cordgrass, is well adapted to frequent inundation by salt water and out-competes other plants to dominate a marsh vista lush green in summer and golden brown in the fall and winter—the reason Georgia's coast is often referred to as the "Golden Isles."

Although most coastal residents once considered marshes nothing more than pestilential swamps, modern ecological research reveals that salt marshes are among the most biologically productive ecosystems in the world, surpassing even the tropical rain forest. A casual look at the marsh immediately suggests that the abundant *Spartina* must be a major contributor. Less obvious are myriad species of microscopic algae living in the water and on the surface of the marsh mud that provide the foundation for the marsh food web. This complex food web supports organisms from those small enough to live between sand grains all the way up to large marine mammals, such as dolphins and manatees. Migratory and resident birds swarm the tidal creeks and mudbanks searching for prey. Humans, too, have learned to value the marsh for its role as a nursery for commercially important seafood species including shrimp, crabs, menhaden, and a wide variety of sport fish.

Where the Altamaha finally meets the Atlantic, we see great forces at work. For centuries, the Altamaha has delivered to the coast enormous quantities of sand derived from hundreds of miles upstream in Georgia's interior. Equally relentless, the ocean has molded that sediment into islands we now call Sapelo, Little St. Simons, Sea Island, St. Simons, and Jekyll. These barrier islands help protect the salt marsh from the full force of ocean waves. If we journey seaward from the marshy western margin of the larger islands, we enter a maritime forest adapted to withstand not only the distinctive climate so close to the coast but also salt spray and the extreme disturbances wrought by hurricanes. The dense, shrubby understory studded with palmettos and yaupon hollies gives way to a canopy dominated by majestic live oaks lending a sense of stability and permanence. But on a barrier island, sand is moving all the time, at all scales, and not just at the beach. Rather, we should think more broadly about what ecologists call the sand-sharing system—the complex interchange of sand between coastal dunes, beaches, and offshore sandbars—all continually adjusting to tides, winds, and currents. On the Georgia coast, where the dominant wind direction is from the northeast, wind and water currents move sand steadily from north to south, building a progression of sand spits at the southern ends of the islands. Cataclysmic storms can move vast quantities of sand virtually overnight, and the Coastal Plain's ancient marine terraces remind us how ephemeral this landscape truly is.

Currently only four of Georgia's barrier islands are accessible by car; three of them, St. Simons, Jekyll, and Sea Island, owe their existence to the Altamaha. Because of their accessibility, these islands attract thousands of tourists each year and are the places where many Georgians first learned to love coastal marshes, the maritime forest, and the seashore. But there are other islands at the Altamaha's mouth—the "bird islands" of Wolf, Egg, and Little Egg. Accounting for less than one percent of Georgia's beachfront, these dynamic islands, sand shoals, and spits bear scant vegetation. Instead, the crop they grow is birds; as many as 18,000 royal terns, 6,000 brown pelicans, 600 Sandwich terns, and 100 gull-billed terns nest there each summer. While some of the Altamaha's seabirds and shorebirds remain close to where they hatch, others range far away. Distinguished by their plump black-and-white bodies and orange, clownlike beaks, juvenile American oystercatchers banded at the mouth of the Altamaha have been sighted nesting as far away as Nantucket Island. The Altamaha islands also serve as wintering grounds and vital stopover points at which migrating birds rest and refuel. In the spring whimbrels arrive from South America and, using their long, downcurved beaks, gobble up fiddler crabs in the salt marsh by day, then rest at night on the treeless bird islands, safe from marauding owls. After five weeks in which the whimbrels double their weight, they lift off for breeding grounds in the Canadian tundra. In September, as many as 15,000 dove-sized red knots descend on the Altamaha's bird islands—their only stop on the East Coast—to gorge on small surf clams before resuming their southward flight, some to winter as far away as Tierra del Fuego along the southern tip of South America. Through their dispersal and migration, these birds stitch the Altamaha into the ecological web not only for Georgia but across the western hemisphere.

Although this unique river system supports an impressive array of plants and animals and diverse habitats, it is far from untouched by humans. Only five years after General James Oglethorpe founded the colony of Georgia in 1733, settlers on the Altamaha established timber and sawmilling operations at Darien. Less than thirty years later, the colony annually exported nearly a quarter-million board feet of pine lumber as well as shingles and barrel staves, most of it cut from the Altamaha's banks. Although it would be pine and cypress that would dominate the river's timber trade for the next 250 years, live oak had a special place among the Altamaha's forest products. Much of the live-oak timber used to build the warship USS *Constitution* in 1794 was cut from St. Simons Island at the mouth of the Altamaha.

The advent of steam-powered sawmills expanded the exploitation of "forests of the finest timber," as the Darien *Gazette* described them in 1822. By 1902, steam-powered cargo ships calling at Darien could stow up to three million board feet of lumber in a ship's hold, the equivalent of all the merchantable timber that could be cut on fifteen hundred acres. Each day as many as eighty-five log rafts composed of up to 100,000 board feet of timber floated downstream on the Altamaha to Darien. Here they were disassembled for milling and export. In that same 1822 article, the Darien newspaper had called the forests "inexhaustible," but the trees were being cut at a clearly unsustainable rate.

Deforestation and the establishment of inland cotton agriculture had an additional impact on the river. Loss of forest cover and erosion of topsoil changed water quality on the Altamaha, an effect that continues to this day. Sir Charles Lyell, an eminent British geologist traveling through Georgia in 1846, wrote:

> Formerly, even during floods, the Altamaha was transparent, or only stained of a darker color by decayed vegetable matter, like some streams in Europe which flow out of peat mosses. So late as 1841, a resident here could distinguish on which of the two branches of the Altamaha, the Oconee or Ocmulgee, a freshet had occurred, for the lands in the upper country, drained by one of these (the Oconee) had already been partially cleared and cultivated, so that that tributary sent down a copious supply of red mud, while the other (the Ocmulgee) remained clear, though swollen. But no sooner than the Indians had been driven out, and the woods of their old hunting-grounds begun to give way before the ax of the new settler, than the Ocmulgee also became turbid.

Thus the Altamaha changed from a blackwater river that ran as clear and dark as iced tea to one with turbid, silt-laden waters, carrying Georgia's topsoil legacy to the sea.

Unfortunately, impaired water quality, poor land-use practices, and deforestation remain issues for the Altamaha today just as they were in Lyell's time, and new threats have arisen to challenge the continuing integrity of the river's ecosystem. Operation of dams, increasing agricultural use of groundwater, industrial discharges into the river, use of water for thermoelectric power, and the water and land demands to support unprecedented population growth along the coast all have impacts on the Altamaha that must be addressed to protect its irreplaceable assets.

The Altamaha's bottomland hardwood forests increasingly are falling before the saw and the wholesale chipping of bald cypress stands into garden mulch destroys valuable habitat. Careless tree-harvesting and associated road-building activities lead to increased runoff and sedimentation that impair water quality. Fertilizers and herbicides used improperly in forest management make their way into streams and rivers with unintended impacts on nontarget species.

The explosive growth of commercial and residential development in the coastal zone has exacerbated wastewater and storm water effects on the lower Altamaha. Rainfall that otherwise naturally soaked into the earth now cascades from roofs, parking lots, and other impervious surfaces to carry contaminants into the river and marshes. Habitats are fragmented or lost altogether as land is cleared for housing and commercial development.

In the past, because the Altamaha basin had a relatively low population density, the river easily met human water needs. But increasing population growth, industrial demand, and the rise of irrigated agriculture combine to stress both its surface and groundwater supplies. Although the contribution may be hard to see, groundwater plays a critical role in maintaining stream flow in the Altamaha during dry seasons and droughts. As human demands increase on both sources of water, excessive withdrawals leave normally submerged areas like sand shoals high and dry, no longer capable of sustaining aquatic organisms that depend on them. When the river flow rate falls, salt water penetrates farther upstream, stressing organisms not adapted to saline conditions, and ultimately altering habitat if continuing long enough. Even salt-tolerant *Spartina* can be stressed by low-flow conditions and elevated salinities as was seen in the alarming dieback of coastal marsh during the severe drought of 2001–2002.

With increasing commercial and residential growth also come increasing demands for electricity, which has marked water impacts of its own. Most Georgians do not realize that thermoelectric generating plants, both coal-fired and nuclear, account for more than half of the water withdrawn each year from Georgia's rivers, primarily for use in their cooling systems. The Plant Hatch nuclear facility near Baxley is responsible for the largest withdrawal of surface water from the Altamaha. Unlike in other cooling system designs, which can return much of the withdrawn water to the river, over half of the water used for cooling purposes at Plant Hatch turns into water vapor, with no guarantee that it will ever return to replenish the river in the form of rain.

Another factor affecting both water quantity and quality is the construction and operation of dams. Although the Altamaha itself remains free-flowing, dams upstream on both the Oconee and Ocmulgee Rivers have a variety of impacts far downstream. Historically, great runs of fish such as shad, striped bass, and sturgeon made their way far inland to spawn in shallow rivers and streams high in the watershed. Both the Indians and early European settlers exploited these seasonal runs as valuable food sources on the frontier. Beginning in the early twentieth century, however, major dam-building projects cut off access to these vital spawning grounds. Today, dams remain an obstacle to restoring the populations of these fish to anything approaching their former abundance. Fish and other aquatic animals must cope with additional changes to the natural system that dams produce. Water released during hydropower generation usually differs from the natural temperature of water in an undammed river. And instead of the natural syncopation of the river's flow over the seasons, dam operations substitute the on-off, on-off pulse of a

Even in the early 1920s, the Oconee River supported an impressive population of migratory fish. This photo shows four Atlantic sturgeon, with a combined weight of 655 pounds, caught at Milledgeville. Photograph by Eberhardt's Studio, courtesy of Georgia Archives, Vanishing Georgia Collection, BAL142.

robot, creating a profoundly different flow regime than the one to which the animals and plants living in and near the river are adapted. Even the coast is affected by dams. Scientists attribute the increased erosion rate there to the multitude of large and small reservoirs and dams constructed since the mid-twentieth century, which trap sediment that otherwise would have replenished the marshes and barrier islands.

Recognizing the Altamaha's unique attributes and the threats it faces, federal and state agencies, nonprofit conservation groups, and private citizens have worked to protect large tracts of the Altamaha River basin. Currently, tens of thousands of acres receive varying degrees of protection in a mosaic of wildlife refuges, wildlife management areas, state and county parks, management agreements with industry, conservation easements, and private reserves. Two of Georgia's most innovative environmental laws, the Marshlands Protection Act of 1970 and the Shore Protection Act of 1981, not only set the pace for the protection of the Altamaha's salt marsh and seashore, but also have served as a model for laws elsewhere in the United States. But laws alone are not enough. Citizens must be vigilant to ensure that agencies continue to enforce laws protecting the lands and waters that all Georgians hold in common, now and into the future. We also need a more holistic way to analyze how our decisions affect the Altamaha ecosystem and the services it provides, because it's not just about saving wild nature, it's also about saving ourselves.

Realizing how much we owe to the Altamaha, many of us have a deep bond with the river. Eight generations of my family have lived in the upper reaches of the Altamaha basin, on the Ocmulgee River near Macon. As a child I stood on a bluff overlooking the Ocmulgee bottomlands and thought it as wild and impenetrable as the Amazon jungle. I longed to accompany the Macon Motorboat Club on its annual summertime trip from Macon down the Ocmulgee and the Altamaha, all the way to the Atlantic. I envisioned sandbars with alligators at every turn, all the way to the sea.

Without a boat, all I could rely on was my imagination. In the intervening years, I studied geology, biology, law, and philosophy to better understand this beloved place. While others know the Altamaha more intimately than I ever will, I know this: I've found myself stuck knee-deep in its marsh mud and laughed at the fiddler crabs beckoning me with their dominant claw to stay awhile. I've stood transfixed as sunrise-pink roseate spoonbills passed overhead to settle in the marsh before me. I've walked the Altamaha's sand shoals, reading the ripple marks. I've searched its barrier beaches for the cast-off cones built grain by grain by sea worms, each fragment of quartz sand as carefully fitted as a pane in a stained-glass window.

I also know this: Last winter James Holland took me to visit the giant cypress. As I waded through the shallow swamp water, my steps released plant oils from the decaying vegetation on the bottom, marking my trail with pools of pastel iridescence. At the record holder, I measured the tree not with a forester's tape but with increments of my arm span, its raveling bark silky against my cheek and fingertips. Nearby stands another giant cypress, this one bearing a deep incision from a crosscut saw. In the early 1900s when the timbermen came to this tract, two sawyers began and then abandoned their work. Like a pilgrim, I touched the wound in the living tree and wondered if they had decided the tree was too big to be hauled out of the swamp. Or was it something else—a reprieve, a moment of grace? The motives of these anonymous men are unknowable. But just as they held the fate of one tree in their hands a century ago, I know we hold the Altamaha in our hands—all it is and all it can be—for those yet to come.

Our River's Keeper

JANISSE RAY

James Holland.
Photograph by Nancy Marshall.

Taking pictures was the last thing on James Holland's mind when he was a kid navigating the streets of Cochran, Georgia. He never imagined the man he would become, the one who would spend years roaming the Altamaha River by boat and on foot, peering through a lens.

First of all, James's family didn't own a camera, and second, more urgent concerns occupied him. Most of his time was spent in a wild abandon that was at once a response to his circumstances, a reflection of his personality, and a search for meaning in a world more roller coaster than carousel, more crumb than cake.

Holland's life was not picture-book. He was born James Rufus Holland to James Quincy Holland and Bessie Alma Holland in a sharecropper cabin in the south Georgia town of Nashville. Somewhere along the way, James Quincy and Bessie's love went awry.

"I grew up in a broken home," says James.

He was six or seven years old when his parents' marriage fell apart, helped along by heavy drinking and infidelity. James and his brother, four-year-old Quincy Lee, found themselves fatherless and motherless and in dire need of a roof over their heads.

A Macon couple, childless friends of the family, adopted Quincy Lee. James was taken in and adopted by his father's unmarried brother, Baxter, although all the aunts and uncles pitched in to help. Not long after, Baxter married the loving woman who would become James's Aunt Charline, and she arrived to the union with a nine-year-old daughter, Alice Faye.

"Faye and I were raised together," says James. "That's how she came to call me her brother."

"He was JR to everybody," Faye told me recently when I called her to talk about James's childhood. "I always thought he was a wonderful person. He would watch out for me." In fact, James reminds her of her late father, whom she describes as a strong, stern, honest man who never drank and never incurred debt. His only vice was the occasional game of setback or pool. "JR has some of the same mannerisms as my daddy," she said. "When he finishes eating and sits back from the table, he folds his arms. My daddy did that exactly. In fact, he looks more like my daddy than he did his own father."

"Our daddy was afraid we would go hungry," Faye continued, "so we spent half our lives picking, shelling, peeling, washing jars, and doing anything you associate with putting food by. When he died, he left five freezers full. He worked like a dog for that food," she said.

"And he fought for orphans," she added. He'd been an orphan himself. "The only time I ever remember him standing up at church was when they wanted to quit sending money to the Georgia Baptist Children's Home."

James's was a jumbled-up childhood, Faye said, "and not very happy." But when she speaks of unhappiness, she is speaking of his childhood as a whole, for there were many wonderful days as well. One of Faye's fondest memories is of one day she and James were sent to their backyard to shell a hamper of peas. It was a long job and the children got tired, so James started sneaking handfuls of whole, unshelled peas in with the hulls. "Mama knew how many peas a bushel would produce, so she knew something was wrong," Faye said, laughing delightedly. "We got into real trouble on that one. We had to shell all the hulls over, empty ones and full ones."

IT IS 2010 and I am riding around Cochran with James, exploring the sights of his childhood. I am eager to see the places that fired in James a love of nature and all things beautiful and the desire to protect them and document them.

It's a beautiful sunny day during a fall in which the weather conditions have compounded perfectly to create incredible, stunning foliage. The trees are fiery, some a deep red, some a transcendental yellow. Cochran, population forty-five hundred, is the seat of Bleckley County and home to Middle Georgia College, which is known as the oldest two-year college in the United States but is transitioning to a four-year, and is now turning out classes of young graduates. The town has its struggles, that much is clear. The downtown area is aiming for revitalization and making admirable progress but hasn't quite succeeded, embedded as it is within half circles of low-income neighborhoods. Young people hanging out on porches and corners in the middle of the day demonstrate that there's some unemployment.

We turn down Ash Street and slow before a small, jaded, pine house. "I came there in 1948," he says. "That's where I spent my third-, fourth-, and fifth-grade years. This street, of course, was dirt back then." Someone else lives in the house now, so we don't stop.

"Wendell Berryhill lived right over there," James says, pointing. "We lived side by side with a street between us. He's my good friend to this day."

"I know you're wondering," James says, "How did I get from an orphan to somebody who loved a camera, especially with all that happened in between?"

"I'm trying to figure it out," I say.

"It had to do with that big ditch over there," James says, "and the time Wendell and I spent in it. We called it the Big Ditch. It's a main tributary of Jordan [pronounced Jerdan] Creek. That was my escape. I played in Hill's Pond too," says James. "That's where I learned to catch all those bluegills.

"If I could get outside, I was outside," James says. "I skipped school to be outside." He remembers one particular day. "I was playing hooky from school, and my daddy came down to Frog Dyke's pond looking for me."

"When you say 'daddy,' do you mean Uncle Baxter?" I interrupt him.

"No, I mean my real daddy. He was back in Cochran by then. He came driving slow, looking for me. I ran like hell. But my father hollered 'JR' and I couldn't disobey my dad. I stopped and called, 'Sir?' I was way across a field and I could have got away. But I didn't. I got my ass caught."

"What did he do with you?"

"Made me go to school."

Just down the street James points out a little black building. "That was a store," he says, "where I bought penny candy." We travel next through a section of town called Happy Hill, past vacant lots that once held a strip of row houses belonging to Hill Timber Company. All that's gone now.

James cruises through the intersection of Cherry and Third, where he also lived awhile. That house is gone too, in its place the offices of AT&T. He turns suddenly and noses down a narrow lane. "Mrs. Mamie Perry ran a hamburger shop in this alley," James says. "She sold burgers for a nickel." Now the hamburger shop is gone, and in front of us are a Dumpster, the brick back walls of stores, and a parking lot.

Lanfair Insurance is in the old movie theater. A movie cost eleven cents, James says. "I didn't have eleven cents very often." What pocket change he had he earned by mowing the lawns of the well-to-do, using a push mower, fifty cents a lawn. The pool hall is now a shop called the Music Box. "When I got big enough—not old enough—I did a lot of hanging out in there," James says. "Just down the street was a beer joint. They sold draft beer in half-gallon jugs. I was thirteen or fourteen when I began trying it out."

By then James had advanced from grass mowing to cotton picking, which gave him a smidgen of spending money. "My knees still hurt every time I ride by a cotton field," he says. "I won't ever have to pick cotton again."

I FIRST MET JAMES in late 1998. He'd been a blue crabber for a couple of decades and he had been observing his catch, and thus his livelihood, decline year after year. "I started looking around, to find out what the hell happened, researching and digging deeper," he said. The state imposed bag limits on crabbers in an attempt to keep crab populations stable, but the numbers declined anyway. "The crash was something we couldn't control ourselves," said Holland, "and I'll tell you, over the last few years when I was a commercial fisherman, we did everything under the sun to save the blue crab fishery and we failed miserably."

As James sank deeper into the science, by reading and attending meetings, he learned to connect the devastation of a fishery with the quality of water upstream. Basically three things were happening to ruin water quality: the degradation of rivers and estuaries; the destruction and development of swamps and marshes; and logging in both wetlands and uplands.

Holland had begun to attend meetings in which the state Environmental Protection Division of the Department of Natural Resources was present. He'd had to learn their lingo—long, scientific, technical, academic, political words, and lots of initials. EPD, DNR, PSP, OVC, SMZ, MOA, BMP. He educated himself on the detrimental effects of water pollution and increased salinity by talking to professors and scientists. He had to learn it all, he who had never had a chance to go to college, who had known nothing except hard work all his life.

As a concerned and increasingly savvy crabber, Holland began raising his hand in meetings. "They started to hate James Holland," he said. "I wanted answers. Sometimes I would hold my hand up and they would go around a time or two before they called on me. I proved that development as much as two miles away affected mussel beds. I never could get anybody from DNR to go with me, so I could show them."

His greatest complaint was governmental complacency—as a new activist, he discovered that state and federal agencies served politicians and industry for the most part, not their mission statements or their human communities. Frustrated at government's lack of action to stem pollution, faced with ever-worsening water quality along the coast, and having fallen victim to the destruction of his livelihood, James decided that he had to do something.

Holland's home watershed was the Satilla, to the south of the Altamaha, at whose mouth he crabbed. Originally he intended to tackle the cleanup of that river. But he knew he could touch more people on the Altamaha: with headwaters that brush the north Georgia mountains, the river drains a quarter of the state's landmass and helps supply the urban centers of Atlanta, Athens, Macon, and Milledgeville.

The state had just released a river basin management plan for the Altamaha in 1998, and most residents didn't like it because it wasn't far-reaching enough. Coastal fisheries relied on clean rivers flowing into them; towns and counties in the watershed sought tourism and an influx of retirees, both of which depended on scenery. The river needed protecting.

I attended the DNR public meeting held in Jesup in conjunction with the release of the management plan. It was December of 1998. A tall, muscular man with a no-nonsense manner occupied a seat on the front row, asking a number of difficult questions, big words and acronyms I didn't understand rolling off his tongue. That was James.

"Used to," he said, "I could harvest 1,500 to 1,800 pounds of crabs off a hundred traps in a day. Now, on a good day, I'm lucky to get 160 pounds. I'm watching my way of living going down the drain. What do you plan to do about it?"

A month later I was in the port town of Darien meeting with concerned citizenry—fishers, biologists, business owners, environmentalists. James Holland led the meeting. Little did I know a group had been assembling for a while in the office of Christi Lambert, who directed the Nature Conservancy's (TNC) Darien office and whose job in large part was creating a bioreserve—landscape-wide protection—in the corridor of the Altamaha River. TNC was working on the land; who was watching over the water?

In January 1999 we founded the Altamaha Riverkeeper (ARK) organization, which was based on the Hudson Riverkeeper model and was the nation's twenty-sixth group to join the Waterkeeper Alliance. We elected James Holland our first Riverkeeper.

I regularly worked with James over the course of my tenure as a board member and also in the years that followed, when I remained a friend and supporter of the organization. I trained with him in stream monitoring, visited sites of devastation, and attended meetings. But for every small thing that I did, James did a hundred thousand. For every sign of devastation I saw, he saw a million. I got to know him well and was pleased to follow his progress as an organizer and an environmental advocate. He was a master at banding people together. He was everywhere at once, and his story, his transformation, galvanized people to support the cause, which was to clean up a major American watershed.

One of the first acts of the Altamaha Riverkeeper was to set up a toll-free hotline, 1-877-ALTAMAH, so citizens could report problems easily, anonymously, and without charge. "I would like to stress," James wrote in an April 2000 newsletter, "that ARK is striving not only to meet or exceed its obligation to the watershed and our membership, but to greatly enhance the waters that most of us love to fish, boat, kayak, canoe, swim, or take our children out for a day on the water. We desperately need your eyes and ears to improve the work that we do."

Within a couple of months of its inception, ARK listed 125 members. The number steadily grew.

James traveled up and down the four rivers of the basin, responding to citizen reports, testing waters with his kits, and confronting polluters large and small. On his way to investigate an unlicensed dam he might encounter multiple acts of violence against the fragile ecosystems of the basin. He used the phone constantly to organize, inspire, and educate. He and Riverkeeper staff, including Executive Director Deborah Sheppard, a dependable and shrewd veteran of environmental work who almost magically raised funds, sent out quarterly newsletters peppered with telling photographs. The damage to the watershed was visible and undeniable:

- Road construction in wetlands with no silt fences in place.
- Hawkinsville, Georgia, South Water Pollution Control Plant discharging blue-colored water into the Ocmulgee.
- High fecal coliform counts in Jordan Creek.
- A landowner destroying hardwood wetlands on Walton Creek.
- Improper stream crossing during timber operations; deep rutting.
- A beheaded alligator floating belly up.
- A Glenwood, Georgia, oxidation pond expelling water with high fecal coliform levels.
- Dead cows from the Hawkinsville livestock barn buried near the Ocmulgee floodplain.
- Carcasses of wild hogs, deer, and goats dumped in Herrin Canal, McIntosh County.
- Big fish kill at Big Sandy Creek and Myrick's Mill Pond due to a kaolin clay processor spilling too much alum.
- Salt marsh destruction due to road building, housing developments, and dock construction.

The pictures hit me—would hit anyone—in the gut. One was of a partially buried cow. One was the telltale purple trail of the plume from the Rayonier paper mill's discharge pipe, near Jesup, where fifty million gallons of effluent daily ruin the color and odor of the river. James photographed a hollow tupelo tree filled with human garbage, an aerial view of the Hatch nuclear plant near Baxley, Rayonier's waste ponds, and toilet paper and human feces floating in a creek.

The worst pictures, for me, were those that captured the true costs of logging our Georgia landscapes. Not only were the pine flatwoods vanishing, entire cypress swamps were being clear-cut, and trees were cut up to the banks of rivers. At that time the climate was staggering in and out of drought, and in the periods of dryness, loggers would creep farther into wetlands. Wetlands connected to rivers, even if they are on private property, are classified as "waters of the state"; the water is owned by the state but the land beneath it is not. What are known as isolated wetlands, meaning not connected directly to a waterway, receive no protection (although as James makes clear to me, all wetlands are connected to rivers, directly or indirectly). Logging in wetlands was undeniably unethical but it was not illegal, and in any case, who was watching? Who was caring? Who was regulating?

JAMES HOLLAND learned as a kid how to be a fighter.

"JR was mean," he says of his childhood self. "I grew up defending my*self*. I also grew up being cynical. I knew which side of the railroad tracks I belonged on. Some people wouldn't even want me at their house. After going through that, why wouldn't I be cynical?"

In the ninth and tenth grades, James played two seasons of football for the Cochran High Royals, as offensive guard and defensive end. In a game against Eastman, a rival player kicked James in the head.

"You hit me, I'm going to hit you back," James said to me.

His opponent had not bargained for a kid who'd won the district shot put championship throwing the twelve-pound shot. After James picked himself up, he punched the Eastman player between the eyes. One can only guess what happened to the player, but James broke two of his knuckles.

Things went downhill from there. James's grades had begun to drop when he was in eighth grade, mainly because, as he said, "I couldn't stand being closed in all the time. I wanted to be at the creek or somewhere. Not in school." In the middle of his tenth-grade year he dropped out of high school and went to work at the Sinclair Service Station, pumping gas.

"I don't want to talk about all the crazy stuff I did," he said.

James turned seventeen on December 1, and the following March 24 he entered military service.

"I was gung-ho, period," he said. "The Marines were supposed to be the biggest and baddest. So was I. Little did I know they were bigger and badder than I was."

He was sent to Okinawa, Japan, where he met Sumiko Sakamoto, a petite, dark-eyed beauty with black hair so long it touched her legs, whom he married on May 21, 1964, and to whom he has remained married, now with three children and many grandchildren between them.

"The United States Marine Corps played a critical role in making me what I am today," James told me. "It taught me self-esteem and self-discipline and to never say 'No' or 'Can't do' until it is proven that it can't be done. No matter how high the hurdles or the number of obstacles in your way, you can and will win when you fight for your beliefs."

Freshly out of the Marine Corps, James took a dangerous and dirty job in a kaolin processing plant near Macon. In the Marines James had been trained as a cook, so he put in an application at Warner Robins Air Force Base, where he was hired as a food services supervisor. For over a decade James would work for Ira Gelber Food Services in a job that would require his family to relocate numerous times. When the contract ended at Warner Robins, James was sent to the Naval Air Station near Brunswick, which has become Homeland Security's Federal Law Enforcement Training Center, now with a zip code and town—Glynco—all to itself. After the food service company lost that contract, James transferred to a naval air facility in Key West.

During his tenure in Key West something changed for James. He worked a second job on his days off, Wednesday afternoons and Saturdays, for a lobsterman, a Mr. Potts. From Mr. Potts James learned how to trap lobsters and crabs, as well as how to run a boat and navigate. He and his family spent fifteen months in Key West.

Then it was on to the naval shipyard in Portsmouth, Virginia, then a dining facility in Norfolk, then back to Glynco when the company recaptured the food contract. In this sojourn in Brunswick James and Sumiko bought a twelve-foot johnboat with a 7.5-horsepower motor. The Navy closed this facility in 1975 and James went to Fort Benning, which had a "pile of mess halls." After a while the company lost that contract.

In 1977 James and his family decided to move for the third and final time, to Brunswick, Georgia, and to leave the food service industry behind.

"Why did you decide that?" I asked.

"You've asked a good question and I will give you the proper answer," James said. "I'm somewhat of a gambler. Food services had gotten too competitive. I was bidding jobs for the company by then. Everything was based on man-hours—how many man-hours can you operate this facility on for a day? I had to put in bids where I knew the company would be marginally able to do the job. That became increasingly difficult."

That was the long answer. The short answer was this: "If you're not careful," James told me, "the world will take your life away."

When Holland moved his family to Brunswick in 1977, he decided to try his hand at crabbing. "I already had a boat," he said. "I saw guys making a living crabbing. I thought I could do it."

By now James had traded up and owned an eighteen-foot fiberglass tri-hull recreational craft, an Aztec.

"Back then it wasn't hard to outfit a crab boat," he said. "You used your hands for everything." He learned how to make crab traps from a neighbor, Sam Beard. By now the children were older and Sumiko would accompany James on the water, putting in a hard day's work in the sun alongside her husband.

With severance pay from the company, the couple purchased a three-acre lot and moved a new doublewide home onto it. This has remained the home place of James and Sumiko.

The couple was successful at crabbing. They returned most of their profit to the business to help it grow. They saved money and purchased a newer, bigger boat. They bought electrical machinery to increase their productivity. They learned to raise softshell crabs and installed concrete holding tanks on the lawn of their home. A second boat sat in the yard, since James knew that if he missed a day's work, he missed a day's pay—if the first boat needed repairs, James fired up the second.

Life was good. The Hollands enjoyed the freedoms of self-employment, the joys of being outdoors, the healthfulness of hard work, the bounty of the ocean. They enjoyed working on the salt sea, watching the dolphins, pelicans, osprey, manatees, and otters. They loved watching the *Spartina* turn straw-colored in the fall, and then grow back freshly green each spring. They loved piling the holiday table with seafood as the children and grandchildren gathered around.

As the years passed, however, no matter how many improvements the Hollands made in efficiency, their return weakened. Their profit margin narrowed. They were working harder to make less than ever. Something was wrong. That's when James began to talk to others in the industry, such as Robert DeWitt, owner of R&R Seafood. "What is happening?" he asked. Nobody knew but they all wanted to find out.

From there, you know the story.

From the day he signed on as Riverkeeper until the day he retired a decade later with the membership of the organization exceeding one thousand, James Holland devoted his life to an environmental crisis playing out in all our lives. Pollution, withdrawal of groundwater, destruction of natural ecosystems, and global warming, all human-caused, are devastating the ability of our planet to support the human species. He couldn't fix the entire planet or all the oceans, but he could start with one place and do what he could.

Environmental violations kept James on the move—not walking, but running. Many of the investigations turned into actions—letters to landowners, alerts to the Army Corps of Engineers, petitions to the DNR, press releases to media. A busted sewage pipe in Eastman, Georgia, for example, dumped straight into Roach Branch, creating fecal coliform counts upward of two million colonies per one hundred milliliters. Safe contact for humans is two hundred colonies. "It leaked for over three years," James said. "The problem stopped when they fixed the leak."

Altamaha Riverkeeper from the start reported the injuries to governmental agencies in charge of regulating environmental offenses. Sometimes the agencies took action, sometimes they didn't. Formed with the intention of entering the judicial realm if necessary, ARK began to litigate and to win. For example, when ARK sued the city of Cochran for polluting Jordan Creek, a tributary of the Ocmulgee, the U.S. district court found that Cochran had violated its wastewater treatment permit at least 105 times between 1995 and 2001. The city was fined $26,250 and later replaced its outdated oxidation pond with a mechanical treatment plant.

The Riverkeeper was arrested three times during his tenure, mostly for trespassing on private property while collecting water samples, although only once did he actually find himself behind bars. He was served a hot dog and fruit cocktail for lunch. "The only thing worse was C rations," he laughed. "You've got to be tough if you're going to be a Riverkeeper."

Once James mastered Internet communication, he e-mailed to his colleagues pictures of what was happening in the basin. I was on his list and I saw the pictures as they came in. One was a town's sewage pipe, its discharge leaving stringy stuff hanging from vegetation downstream. "Y'all be the judge and the jury," James wrote.

About another discharge point he wrote, "It smells like soap around this pipe. Maybe they call soapy water clean water?"

He sent pictures of the Rayonier paper mill discharge pipe, images of foam and purple water. Years later, after meeting with Rayonier, he sent still more pictures. "Still nasty as all hell," he wrote.

"He took scads of photos of how hideous things were," said Deborah Sheppard, ARK's executive director:

- Sewage spill on Bay Street in Brunswick. Another at East Dublin Elementary.
- Blount's Crossing Road developer destroying wetlands.
- More elevated fecal coliform counts.

- A wetland used as an excavation site for fill dirt. Destruction of isolated wetlands. A wetland being prepared for pine planting.
- Illegal ditching.
- A sheen of diesel fuel across the surface of the Darien River.
- A swimming pool being dug in a salt marsh buffer zone.
- Demolition debris dumped in the floodplain.
- Illegal boat ramps. Illegal docks.
- Wayne County building a road to a new school through hardwood wetlands, without a permit.

No polluter or destroyer could hide from Holland, and when they tried, with beauty strips and backwoods abominations and No Trespassing signs, he took to the air with SouthWings, an environmental flying service that watches over the South. "Shame on the person who did this," he would write, or, "How to ruin a perfect day." And, "He needs to be in jail along with his cohorts and if I can help him get there I will sleep good at night."

Constance Riggins, development director for ARK, wrote a beautiful tribute to James in the Winter 2007 issue of the organization's newsletter:

> James has both a realistic and optimistic vision for the future. His voice quakes when he hears about fish kills. He is heartbroken when he witnesses wetland destruction. He fears soon the small springs and creeks we know will become bare skeletons of what they once were or join the long list of wonderful things that no longer exist. He hates the destruction of trees and wild animals for no reason. Every case of pollution James discovers is the worst, most destructive, and stupidest thing he ever saw, and he will stop it; he won't take no for an answer. . . . James never takes a day off unless it is to go fishing; however, even then many outings turn up more work. It is just as well because he would rather be working to protect our watershed than doing anything else in the world.

In all the years of organizing, the campaign that James remembers most fondly is Gum Swamp on the Ocmulgee. When Eastman proposed a land-spraying plan to dispose of its treated wastewater, in an attempt to solve the Roach Branch problem (fecal coliform from faulty sewage pipes), residents in the area of the spray fields complained to ARK. They were fearful of living with odors and health problems. "I never dreamed those folks would stand up and say no," James said. "I'm not talking activists. I'm talking farmers, housewives. They organized themselves."

"I told them, 'Folks, I can't do this for you. You're going to have to fight this battle yourselves.' I told them that if they fought I could guide them through the governmental processes. And they did. And they won. I could never be more proud of a group of individuals than I am of those folks. Those who have the courage to fight usually win, one way or another."

"My job was education," James said, "up and down the watershed. The hardest lesson for people was how to deal with adversity in their own community. ARK was an instrument in offering guidance."

Holland traveled up and down the river basin talking. He spoke the people's language, addressing garden and Rotary and Lions and hunt clubs and chambers of commerce; college and high school and elementary classes; city councils and county commissions. He went to Atlanta, to Athens, to Dublin, to Brunswick, to Jesup. He went to Hazlehurst, Ludowici, Macon, Lumber City, Everett City.

To Savannah, Baxley, Odum, Gardi. Up and down the river, in and out of the watershed he went. Holland gave hundreds of people tours of the swamps and wetlands of the Altamaha basin.

When Holland stood in front of classrooms of students, he first apologized to them. "My age people have cheated you out of what belongs to you," he said. "We owe you an apology."

Slowly, a people began to reconcile themselves with their landscape, with their home, and with each other. Federal judges were sympathetic to environmental law—which is to say, ethical—and every case that Altamaha Riverkeeper took to court it won. A river began to win.

SOMETIME IN 2003 the editor of *Georgia Outdoor Adventures*, a tabloid based in Perry, asked James if he would contribute an article once in a while, and thus began a beat that lasted until James couldn't keep up with the deadlines because of his workload. This series of essays proved Holland's abilities as a writer, his commitment to nature, and his knowledge of government and policy.

In one, "Are Best Management Practices for Forestry Being Used Properly?" James began: "In the last few years the huge increase of timber harvesting in Georgia's floodplains has increased the probability for mistakes that are detrimental to water quality and fish habitat." He went on to acknowledge that landowners shouldn't be obstructed from harvesting their timber, but he stressed that they should follow the strict guidelines put in place by the Clean Water Act and by Georgia law. The article called for training of loggers in interpreting, understanding, and complying with applicable laws.

In a follow-up article that looked at the enforcement of logging guidelines, James mourned the decline of populations of wood duck and other game species due to loss of nesting and acorn-producing trees. "We are losing this habitat at an alarmingly rapid rate by so-called minor ditching (three feet by three feet ditches) and draining," he wrote.

In another article, titled "Conservation Is Good Economics," James wrote about learning to hunt quail. "I was taught as a young boy never to kill all the birds in a covey, even if I could. It was and I assume still is common practice to leave five or six birds alive in the covey so they can multiply for the future." He warned that degraded habitat in the bottomlands "may very well be costing the state hundreds of thousands of dollars in lost revenue each year from the diminishing populations of ducks and duck hunters" and called for the return of duck hunting to its "glory days."

In another piece, "Georgia's Natural Salt Marsh Systems," James Holland declared that poet Sidney Lanier would be in tears if he could see what we have done to ravage the marshes of Glynn, of which he wrote.

"For even then," James reminded his readers, "[Lanier] wondered what swam beneath the waters on a high tide in these fabulous salt marshes."

Holland knew what swam in the waters. He knew also that many of the creatures were in peril. "Black drum and Virginia croaker were quite common in some of our waters in the past twenty-five years," he wrote. "The list continues with mullet, saltwater catfish, spotted and summer trout. As most fishermen will attest, this is no longer the case.

"There are without question fewer blue crabs, and if you believe the old-time shrimpers (and I do), much less shrimp," he continued. "There is no longer a truly viable oyster, menhaden, Atlantic shad, or Atlantic sturgeon fishery on our coast, especially in the Altamaha watershed."

He then raised the question of whether these declines were caused by natural ecological events or by something else: "Could this be man's killing machine at work?"

IN THE END, the travesty was too much even for even James's calm, rational, Marine-trained mind, too much for his immensely capacious heart. He couldn't keep focusing on tragedies. Transformation came one fine day in the form of a camera.

"My life from that point escalated into something else," he said.

Deborah Sheppard remembers almost the exact moment. "We were doing a cleanup at the Wayne County Landing," she recounts. "We had been joking with James about tree huggers. He said he definitely wasn't a tree hugger. Then I looked and I saw him taking a picture of a zebra butterfly on a lantana. Wow, I thought. That was the first time I noticed him doing that.

"I think it was butterflies that hooked him and moved him toward pictures," she said. Brad Winn, a biologist with the DNR, saw the pictures and said to James, "If you're going to do this, you need a good camera." At that time, around 2004, James was deeply interested in wildlife photography but knew little about it. He was using a small, simple 35-mm Sony camera. "Brad Winn and Jeannie Lewis introduced me to the Canon EOS30D," James said. An education grant from the Nongame Section of the DNR provided that equipment.

"That camera and a long lens [a 200–400-mm zoom] really put me in photography," James said. "I could do so much better."

Now James began to send pictures of beautiful things, wild things, rare things, endangered things. Tiger swallowtail butterflies, wood storks constructing nests, raccoons washing food, water hyacinth, gulf fritillaries, roseate spoonbills, sunning alligators, four wood ducklings on a log.

"Even when I was seeing the degradation, I saw that beauty was still there," he said. "I found out that the most beautiful flowers on God's earth are around wetlands."

Sheppard continued, "One day James and I and Billie Jo Parker, the first Coastkeeper, were on a boat trip. She was driving and he was shooting pictures of a great blue heron. When I saw the picture, I could see the wind moving the feathers. That was the 'oh my God' moment."

The image is of a great blue heron crossing the Darien River, one of the delta outlets of the Altamaha, from right to left. The heron seems close, only feet away from the lens. Its body is the characteristic S-shape of a heron in flight, legs outstretched behind. In the instant of the image being snapped, the wings are uptilted. The farther wing is powder-blue layered on the slate-blue of primary feathers, with a coin of orange at the joint. The heron's yellow eye is attentive, comprehending, intense.

"I think that great blue heron photograph may be my favorite," Sheppard said. "After that, I think that photography became the fuel that fed James's fire, motivated him, and opened him up to let others see his soft side."

As time passed, James continued to upgrade his equipment, next, to an EOS40D, then in late 2009 to an EOS7D. He uses two zooms, a 200–400 and a 100–200. Often he goes into the woods with two cameras at the ready, since the shorter lens is better for landscapes. He shuns filters. He owns a wide-angle lens and "a macro I haven't mastered," he said. Instead, he snaps close-ups with the long lens.

I began to collect his photos by saving his e-mails in a digital mail folder. Little did I know, my husband was tucking away his own favorites. And we weren't the only collectors. As James's list of recipients grew, his colleagues recognized something powerful in his photographs. More people than I can recall have revealed to me their stash of Holland's work, splendor that can lift the spirits, brighten a day, invite the outdoors inside, and strike the flint of wonder.

"His eye for beauty is amazing," James's sister, Faye, told me. "This is surprising since his life didn't have much beauty in it. That big galoot can spot a tiny flower and I don't even see what he is looking at, not until he shows me in slow motion." Three of James's prints hang in her living room. "And it's not near enough," she said.

As Constance Riggins wrote so aptly and eloquently, "I will probably never really fully know James Holland. However . . . with each picture of a creature, flower, or bird that he takes and shares, I get to know James a little better."

AS THE YEARS PASSED, James's political influence grew and he was increasingly quoted in news articles and otherwise recognized as the authority he had become. In 2005 the Georgia First Amendment Foundation and the *Atlanta Journal-Constitution* named James Holland a Citizen Hero for his employment of Georgia's Open Records Act, which, as *Georgia Trend* magazine contributor Jerry Grillo wrote, he wields "like an expert swordsman." In acknowledging the award, Deborah Sheppard called Holland a "force of nature."

In 2008, at age sixty-seven, James Holland joined CEOs, college presidents, and statesmen in being named by *Georgia Trend* magazine as one of the 100 Most Influential Georgians. The magazine included him on the list because he is, in the editors' words, "a powerful protector and persuasive spokesman for the environment."

Holland has thoughts aplenty on Georgia's political topography, which in too many instances protects industry, not nature. "I'm a lifelong Georgian," he says. "But sometimes I'm ashamed of the way we do things. We need a change of climate in our political environment," James says. "We have a state political system that doesn't believe that the economy can coexist with the environment. We need to change that climate. Human existence on earth relies on that."

The economy cannot survive without the environment. Neither can humans.

James continues. "So many problems are staring us in the face—pollution, unsustainable logging, the climate crisis. We have people in power who don't believe that humans are the problem," he says. "Until we change that we're going nowhere. Humans are the problem."

Politicians manipulate and direct the regulatory agencies, which in turn make decisions based on political lawlessness and lack of backbone. "Georgia has some very good laws, and they could be very effective if they were enforced," James says. "EPD is too quick to write a consent order to allow a man to keep a buffer. He's just destroyed a bunch of marsh building a long dock to open water so that he can get from his house to his boat. EPD fines him forty-five hundred dollars but they allow him to keep the buffer. Well, that's against the law. There are no after-the-fact buffer variances."

He is speaking both of a specific violator and of dozens of others when he says, "We don't want his money. Put the marsh back the way it was and leave it alone."

James believes that there's never a good reason to violate environmental law. But, he says, the person who doesn't know he or she violated the law should not be treated the same as the person who knowingly violated the law, who knew it was illegal to build a long boat dock through a salt marsh or log a wetland or dump the river full of chemicals that resulted in a fish kill.

I ask James about his modus operandi. "Some environmental groups take some water samples, make some graphs. They educate a few people and make some corrections, but on the whole, what good do they really do? Others do a remarkably good job but are informed by board members as to what they can and can't do. Generally speaking, a group dealing in policy is more inclined to give way to polluters.

"When I was Riverkeeper, there was no blend. I'm harsher. It is either black or white. You were taking care of the problem or you weren't.

"The political environment in Georgia says you lose when you try to hit a happy medium," he says. "If you screwed up, you shouldn't get away with it. You should fix it."

His vision for the Altamaha is to leave an environmental legacy for our children and their children and so on, in hopes that they might get to experience the beauty, intricacy, and adventure that have been ours, and then some.

ON THE DAY James tours me around Cochran, we ride out past the city-limits sign and on to Union Hill Baptist Church, where after all these years he is still registered as a member. We eat lunch at Scott's BBQ, which serves Brunswick stew that contains English peas and spaghetti. We pull into a Citgo station, white with red poles. James's stepmother's brother started this station. James pumps gas and goes inside to pay while I wait in the truck. In his console are two pocket-knives, some pens, mints, a Jekyll Island pass, glasses, and a mobile phone. In the pocket of the door is a book, *Field Manual for Erosion and Sediment Control*. Behind the seat, the king cab is full of photography equipment.

On the ride back toward Brunswick, where James lives, we talk about the art of photography. Suddenly he brakes because he has seen something photogenic: a hardwood bottom glowing with fall colors, leaves the color of persimmons, of dogwood berries, of pumpkins, of gold coins. The awns of sedges are ripe with seeds. Wild cherry leaves hang dry like so many moons against a cobalt sky.

I watch James take pictures, first from the shoulder of the road, and then from the bed of the truck. I worry about a car hitting him. When he reenters the cab he says, "I get a lot of photos off roads like this."

"Did you get the one you wanted?" I ask.

"I hope so. I won't know until I get home."

"Remember to put me on the recipient list," I say.

"The camera ain't been made that'll capture it like it really is," he says. "But in this case, you just let the brilliance of the leaves do the work for you."

Holland has had no formal training as a photographer. "When it comes to cameras and photos," he says, "I'm self-taught." I suppose, then, we could call him a folk photographer except that his photos are so professional, so mesmerizing, so perfectly composed. An e-mail from James with a photo attachment is like a little gift to be opened—I am assured that the photo will take my breath away or astound me or fill me with longing to be deep in the wild.

Beyond that, many of James's works exhibit some deeper quality, a spirituality that is somehow captured in the image. They elicit praise for Creation.

I ask James what advice he had for a beginning nature photographer.

"I don't know anything about shutter speeds," he says. "But if you can get close enough, while respecting the rules of wildlife ethics, you'll get a good photo. I've even sneaked up on wild turkeys." Of course, a powerful lens helps too.

"The photo you want you only see for a split second," he says. "Lining up the perfect moment in the perfect spot twice is hard. As far as wildlife is concerned, birds are especially difficult. They move so much. I've done it enough now, I can get two, three, four shots of one coming.

"My knowledge of wildlife will help me get the photo I want," he says. "If you have the advantage of seeing something coming, you'll get a decent photo."

That said, he admits that "sometimes a single shot is better than four or five."

"Patience is the key," he says. He mentions a photograph he took of a spike buck, and I remember the one he means. "I stood still for a good twenty minutes to get that shot. While I stood there, the spike moved toward me, from two hundred yards away to about forty yards away. The wind was in the right direction and I didn't move.

"If you're quiet, you can actually move some," James continues. "But if the animal lifts that head up, you better be standing still.

"Second," he says, "practice at moving targets. Number three, delay doesn't work. You need a camera without a delay."

James has one more piece of advice. "Think out of the box," he says. Maybe that is what makes his work so captivating. He doesn't come at wildlife photography from a predictable place. He sees nature freshly. His images take the viewer by surprise and also by storm.

HOLLAND HAS BEEN ASKED why he takes so few pictures of humans. "It's not about humans," he says. "Humans can think for themselves. It's the animals that can't think for themselves. For the most part, humans are the problem. If we leave habitat and clean water for the creatures, humans will do well also."

In his heart of hearts, everything James does is for children. "Inner-city children don't know what's in the woods," James says. "How can they know? They can't get out there to see it. They don't have the abilities. How can they know if we don't take it to them? So we take it to them." In pictures.

His next words cause tears to spring to my eyes. "Here, children, here," he says. "Here it is. We're giving it to you in full color."

In our conversations James keeps coming back to the subject of beauty. There's something that he's trying to get me to understand, and I am slow at understanding it. "I grew up with nature," he says, "but it was a different world. In a way, I was in it and didn't see it. Or maybe I saw it and didn't understand what I was seeing. I grew to view and respect nature in ways that I never dreamed of as a child or young adult. Life takes strange twists. Now I understand that me and nature are the same."

AS COLLEAGUES AND FRIENDS, James and I touch base often over the phone, at meetings, and online. One day a couple months after our trip to Cochran, we were confabbing on the phone. James, who had been subpoenaed to Atlanta in a court case involving destruction of salt marsh, recently had watched a video clip featuring Jean-Michel Cousteau, the son of Jacques. Jean-Michel's message had been positive: if we change our ways, the oceans will rebound.

"I've got news for Mr. Cousteau," James said. "What world does he live in? The oceans are gone, really gone."

I was silent and James paused.

"It makes me ask myself, 'What the hell have you been doing all these years?'" He stopped, sighed. "What did we accomplish? We have a state government that could care less about water quality and habitat. It's really sickening."

After a decade as the Altamaha Riverkeeper, the time came for James to step down. The day we toured Cochran, in fact, he was a few months retired.

"Nobody could be as intense as I was and last more than ten years," he said. "I never grew numb but it just got to the point of me asking myself, 'What am I doing? I'm so tired. I don't get to live my life.' In a job like that, you're taking what's coming at you, not what you want."

In the end, maybe that's why James turned to photography. The butterflies and spoonbills weren't coming at him, he was going after them. They didn't care about him. They had no bad news, nothing to fight. They flitted on by.

And yet, on his last official day on the job, he called to tell me that the City of Jesup was dumping raw sewage again, and the mess was visible around the pipe, stringing in the trees. "There were more condoms than you can believe," he said. "Every year they do it. We talk to them and talk to them and they keep doing it."

James's hope in sharing his photographs is that seeing the beauty of a place will inspire in its citizenry the desire to defend and protect it. Think of each butterfly as an orphan child, needing love and care.

Think what you would do.

Think what you *can* do.

Visions of the Altamaha

PHOTOGRAPHS BY JAMES HOLLAND CAPTIONS BY DORINDA G. DALLMEYER

To know the Altamaha, you need to know its beginning. The Ocmulgee and the Oconee rivers are the tributaries that form the basis of the Altamaha. This is my beginning as well, since I grew up in these headwaters.—JAMES HOLLAND

Page 28: Although very common, the eastern tiger swallowtail butterfly (*Papilio glaucus*) nevertheless brightens the landscape wherever it flutters. Males are always tiger-striped in yellow and black, but some females are dark charcoal with even darker stripes. Here the butterfly is feeding on nectar from cerise flowers of the obedient plant (*Physostegia virginiana*), so called because if its flowers are turned left or right, they will hold that position. This spreading perennial plant, which thrives in wet, rich soil, is also known as false dragonhead because of its resemblance to snapdragons.

The Altamaha, formed from the confluence of the Ocmulgee River (*bottom*) and Oconee River (*left*), is crossed by only eight bridges along its 137-mile length. Railroads and highways have taken the place of this once-vital inland waterway.

Because the Altamaha flows across the gently inclined Coastal Plain, the river develops meanders in its floodplain. On the inside of the river bends, eroded material is deposited in sand shoals called point bars. On the outside of the river bends, the river erodes and undercuts the bank until it caves in. The Ocmulgee River flows in from the lower left, the Oconee River from the upper left, and the Altamaha flows east toward the right from the confluence.

The American alligator (*Alligator mississippiensis*) is one of the most recognizable residents of the Altamaha basin. Relentlessly hunted and trapped for its hide and meat, alligators had reached such low numbers by the 1960s that they were placed under federal protection by the authority of the Endangered Species Preservation Act of 1966, the predecessor of the Endangered Species Act of 1973. During two decades of stringent protection, the population rebounded so successfully that the species was removed from the list, and now strictly-controlled hunting and trapping are allowed.

A river makes its presence felt well beyond the main channel. This particular cypress pond connects with the Ocmulgee River approximately a mile away. In a rarely observed event, spawning shellcrackers (*Lepomis microlophus*) make use of the river's warm, slow-moving sloughs for spawning. The fish get their common name from a diet dominated by snails and crustaceans, which the shellcracker crushes using pharyngeal plates in its throat.

The wide-ranging green tree frog (*Hyla cinerea*) is Georgia's official state amphibian. In recent decades its range has expanded north of the Fall Line, perhaps aided by the creation of small farm ponds and reservoirs and by ponds built by a resurgent beaver population.

This stick insect, otherwise well camouflaged as just another dew-covered twig, stands out against these green leaf tips. Stick insects often rock from side to side as they move, either to mimic vegetation swaying in the wind or to help visually discriminate between items in the foreground and the background.

A meticulous nest builder, this female orchard oriole (*Icterus spurius*) weaves a deep, hanging basket of fine grasses between the branches of a red oak where the male stands on guard. Orchard orioles feed mostly on insects and provide beneficial pest control services on rural farm land.

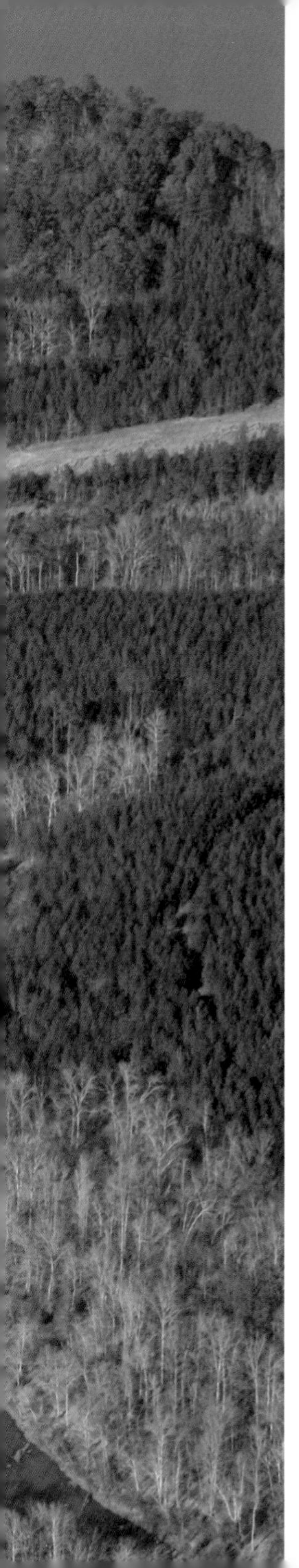

The common green darner (*Anax junius*) is indeed common throughout Georgia anywhere there is calm, fresh water. When plenty of prey is available, it may gather into feeding throngs. In autumn, throngs of migrating common green darners appear at the coast.

This dam on the Oconee River supplies hydropower as well as impounds Lake Sinclair, which serves as a source of cooling water for the Plant Branch coal-fired electric generating plant located on its shores. Dramatic variations in water level due to hydropower releases interfere with the ability of many aquatic organisms to continue to inhabit the river for an extensive stretch downstream.

The Atamasco lily (*Zephyranthes atamasca*) is a sure sign of spring in the rich Piedmont woods of the Ocmulgee and Oconee.

Commonly found along Altamaha tributaries in the Piedmont, the sweet betsy trillium (*Trillium cuneatum)* is an adaptable, vigorous plant preferring rich soil on the uplands. Sweet betsy smells similar to bananas and is pollinated by bees, unlike the many other trillium species that smell like rotten meat to attract flies and beetles.

Having a banana odor similiar to that of the sweet betsy trillium, the rich red-maroon flowers of the spotted trillium (*Trillium maculatum*) rise above leaves mottled with silver and light green. The plant requires good drainage with a rich organic surface layer atop limestone soils developed on river banks and bluffs. Only rarely does a plant produce lemon-yellow flowers, as seen here.

Far more common in the Piedmont and the mountains, the broad-winged hawk (*Buteo platypterus*) is a rare local breeder along the upper reaches of the Altamaha. This hawk feeds primarily on reptiles and amphibians it seizes on the forest floor.

The great crested flycatcher (*Myiarchus crinitus*) is found throughout the Altamaha basin. It spends most of the day high in the tree canopy, where its raucous "*wheep*" gives away its location. A cavity nester, the bird is famous for including shed snakeskins among its nesting material, not to deter snakes, as many old tales say, but more for decoration.

Improper clear-cutting of hardwoods at a bend of the Oconee River in Wilkinson County fouls the river with muddy runoff. Silt consistently tops the list of water pollutants in Georgia. In addition, valuable riverine habitat has been fragmented and may be lost altogether if it is replaced with rows of planted pines as seen in the upper left.

Industrial discharge from a newspaper recycling operation formerly run by SP Newsprint is clearly visible in this aerial view of the Oconee River near Dublin. After the Altamaha Riverkeeper threatened to file suit to stop the discharge of wastewater full of plastic debris, the company entered into a voluntary agreement in 2005 to install new technology to reduce the amount of plastic in its wastewater and to monitor downstream water quality on a regular basis.

Aquatic organisms have evolved over millennia to live within a certain range of conditions, such as seasonal swings in the water temperature of the river. But the relatively recent addition of human-induced stresses, such as chemical pollution, may impair the ability of aquatic organisms to cope with both natural and human-made extremes. Lesions at the base of this bluegill bream's tail are believed to result from water pollution that has weakened the fish, thereby making it more susceptible to diseases prevalent during warm summer months.

The blue grosbeak (*Guiraca caerulea*) prefers open country such as abandoned fields overgrown with briars and openings left by logging operations. A summer resident in the Altamaha basin, the male bird is frequently seen singing from the top of a fence post or bush.

This pair of pileated woodpeckers (*Dryocopus pileatus*) will remain together in their territory year-round. In addition to excavating a nest cavity spacious enough to accommodate the sixth-largest woodpecker in the world, the birds chisel distinctive rectangular gouge marks in their pursuit of wood-boring insects. The pileated's loud "*kuk-kuk-kuk*" call and its selection of particularly resonant limbs for drumming make it a unique part of the Altamaha soundscape. The male bird is the one on the right, with the red cheek patch.

The Altamaha is one of the most ecologically diverse river systems in the world. Its main stem hosts some of the rarest plants and animals in the United States. Included among these are the spinymussel and the largest population of the endangered shortnose sturgeon south of Cape Hatteras, North Carolina. Dicerandra radfordiana, *which has less than three hundred known plants on earth, is part of this most valuable basin.*—JAMES HOLLAND

Giant river cane (*Arundinaria gigantea*) formerly was widespread across the South, where impenetrable canebrakes were the bane of travelers. Historically, this native bamboo probably supported large populations of bison along with other grazers, and settlers took advantage of cane's palatability as cattle forage. Like its Asian bamboo relatives, cane blooms only rarely. Although it may set seeds, few of them survive, so the plant depends primarily on spreading rhizomes below ground.

Widespread along the Altamaha, these Indian pinks (*Spigelia marilandica*) brighten the shadows cast by great hardwoods overhead. Known to the Indians and early settlers as "worm grass," Indian pinks yield a root extract used to rid children of intestinal worms.

The cloudless sulphur butterfly (*Phoebis sennae*) prefers to feed on long, tubular flowers like those of Carolina jasmine (*Gelsemium sempervirens*), a spectacular vine whose flowers festoon the Altamaha's banks in early spring.

The brilliant yellow plumage of the prothonotary warbler (*Protonotaria citrea*) is the source of its common name "swamp canary." A cavity nester, it depends on birds like the downy woodpecker to provide its home. It prefers large, unbroken stands of trees and thus is at particular risk from the loss of bottomland forests to logging and development.

Wendell Berryhill, a lifelong friend of James Holland, hauls in a largemouth bass.

Erosion and transport of sediment cause river meanders to curve so widely that the Altamaha eventually cuts through the base of a loop to shorten its path to the sea. Although it is not yet completely cut off from the river flow by accumulating sand shoals, an oxbow lake is beginning to form on the left side of the vegetated island. Isolated from the rest of the river except during floods, these lakes are famous among the angling community for record largemouth bass and other game fish.

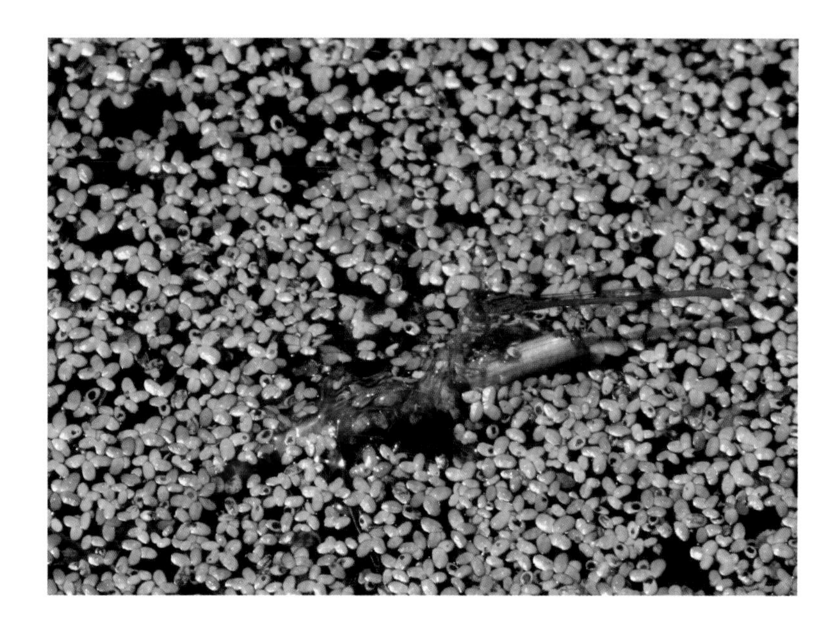

A rare damselfly occurring only in scattered locations, this bright red duckweed firetail (*Telebasis byersi*) rests on its favorite habitat: a duckweed mat floating on still, fresh water. This species of damselfly seldom strays far from either water or duckweed.

The brilliant red cardinal flower (*Lobelia cardinalis*) graces sunny banks along the Altamaha.

The common moorhen (*Gallinula chloropus*), a colorful member of the rail family, prefers freshwater marshes like this pond choked with duckweed. A reluctant flyer, the bird is quite nimble, with long toes on its unwebbed feet enabling it to walk across lily pads and climb through dense reeds. The prominent red projection at the bend of each chick's wing is the first digit, equivalent to the human thumb.

Mysterious sandy circles in the middle of a floodplain track are the work of the redbreast sunfish (*Lepomis auritus*). When the Altamaha floods its banks, fish move out into the floodplain, too, constructing these spawning beds where they and their eggs are less vulnerable to predators. A favorite of Altamaha anglers for many years, the redbreast population has been dramatically reduced by the introduction of flathead catfish native to the lower Great Lakes and the Mississippi River basin.

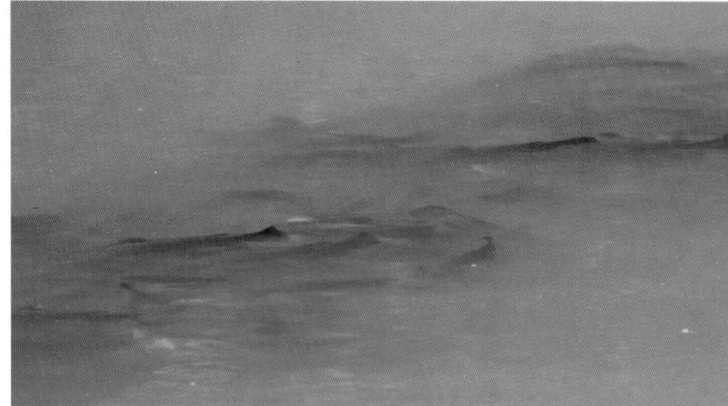

Yellow bullhead catfish (*Ameiurus natalis*) school on the Altamaha floodplain during high water.

The Juvenal's duskywing butterfly (*Erynnis juvenalis*) is associated with oak woods where its caterpillars feed on the leaves. A common butterfly, it nonetheless is difficult to observe because it takes flight quite readily if approached.

The white-leaf leather flower (*Clematis glaucophylla*) is an eye-catching vine whose hot-pink flowers and abundant nectar are particularly attractive to hummingbirds. Although each flower is less than an inch long, they can cover the vines all summer.

A member of the morning glory family, the wild potato vine (*Ipomoea pandurata*) produces tuberous roots weighing up to thirty pounds. Indians dug up these ponderous roots to roast into a starchy foodstuff.

The rare Alabama milkvine (*Matelea alabamensis*) is an unusual trailing member of the milkweed family restricted to sunny openings on forested bluffs at only a few sites along the Altamaha.

A member of the lily family, fly poison (*Amianthium muscaetoxicum*) is aptly named, for the early settlers crushed its bulbs in water mixed with sugar or honey to attract and kill flies.

The yellow fringed orchid (*Platanthera ciliaris*) is found in sunny, wet locations. As wetlands are drained and filled, this orchid and its companion bog plants become increasingly rare.

The overhanging cowl of the hooded pitcherplant (*Sarracenia minor var. minor*) helps to trap insects crawling inside searching for its nectar. Insects attempting to leave may mistake the translucent spots on the hood for openings to the sky and exhaust themselves trying to fly out. Like most insectivorous plants, the pitcherplant depends on insect prey to supply vital nitrogen lacking in boggy soils.

The floating bladderwort (*Utricularia inflata*) is a showy carnivorous plant often found growing in still waters of the Altamaha basin. With the exception of the flower stalk, the plant is submerged, including the inflated leaf stalks that support the plant near the surface. The plant feeds mainly on tiny crustaceans that are caught when they contact trigger hairs surrounding the mouth of the bladders. The bladders quickly expand, sucking the prey inside where it is dissolved to provide vital elements for plant growth.

This "spike" buck whitetail deer (*Odocoileus virginianus*) stands ready to spring into the shrubs bordering a floodplain dirt track. Although the common assumption is that unbranched antlers indicate a young deer, research with captive deer herds indicates that the growth of spike antlers can occur at any age depending on genetics and nutrition. The rich hardwood forests of the Altamaha bottomlands provide plenty of sustenance for these adaptable browsers.

Historically, the wild turkey (*Meleagris gallopavo*) was one of the most abundant game birds in North America. As in the rest of the United States, excessive hunting and habitat loss reduced its numbers in Georgia to a low of 17,000 birds in 1973. Scientifically based management and support by conservation groups has helped the wild turkey population rebound to an estimated 300,000 statewide.

These wild azalea blossoms (*Rhododendron canescens*) provide nectar not only for insects but also for hummingbirds migrating through the Altamaha corridor.

Bartram's rose-gentian (*Sabatia bartramii*) is named in honor of William Bartram, a pioneering naturalist who explored the Altamaha on many occasions during his visit to the frontier South in 1773–1777. On the south side of the Altamaha, Bartram described seeing "extensive green savannas checquered with the incarnate *Chironia pulcherrima*," which modern botanists believe may well be this species that now bears his name.

The barred owl (*Strix varia*) inhabits wooded swamps all along the Altamaha, where it nests almost exclusively in hollow trees. Barred owls are quite vocal, especially during the breeding season. Hearing a chorus of owls scattered up and down the river trading their signature call "*Who cooks for you? Who cooks for you all*?" is an unforgettable experience.

A small yellow-bellied slider (*Trachemys scripta*) perches atop a much larger companion to take in the sun.

The Plant Hatch nuclear power facility near Baxley is responsible for the largest withdrawal of surface water on the Altamaha. Unlike in other cooling system designs, which can return much of the withdrawn water to the river, over half the water used for cooling purposes at Plant Hatch turns into water vapor, with no guarantee that it will ever return to replenish the river in the form of rain.

An uncommon Georgia dragonfly, the stripe-winged baskettail (*Epitheca costalis*) is found along slow stretches of the Altamaha and its backwaters. Males lack the brown stripes this emerging female displays.

The eastern comma butterfly (*Polygonia comma*) is seen in the Altamaha's deciduous woods, especially those near swamps and marshes. Here the butterfly is using its proboscis to suck up minerals and salt from wet soil, a behavior known as "puddling." The eastern comma prefers sap and rotten fruit to feeding on flowers.

Tulip poplars (*Liriodendron tulipifera*) reach great size in the Altamaha bottomlands. A member of the magnolia family, the pale green and orange flowers brim with nectar in the spring.

The diminutive blue-gray gnatcatcher (*Polioptila caerulea*) is a year-round resident in the Altamaha floodplain forests. Always in motion, this tiny bird inspects leaves for insect prey, occasionally sallying out to capture them in flight. It announces itself with soft mews and is quite curious about human activity.

This adult male eastern fence lizard (*Sceloporus undulatus*) shows off its intensely blue belly patches during the spring breeding season. Fence lizards spend much of their time in trees and are more abundant in open forests than in swamps. Against tree bark, their color and rough texture help camouflage them from predators.

Red advertises the availability of nectar, ripe fruits, and seeds to animals who will go on to disperse the plant's offspring over a wide area. *Above left*, nectar deep in the funnel-shaped flowers of trumpet creeper (*Campsis radicans*) is a favorite food source for butterflies and hummingbirds throughout the summer. *Below left*, the pyramid magnolia (*Magnolia pyramidata*) is one of the rarer magnolias, found at only a few sites along the Altamaha. *Below right*, the mock strawberry (*Duchesnea indica*) is a non-native plant with little to commend it, for it spreads rampantly and its berries have no taste. *Above right*, during most of the growing season, heart's-a-bustin (*Euonymus americanus*) is easy to overlook. In the fall, however, the plant is adorned with red-purple husks bursting with brilliant orange seeds, a prized delicacy of deer.

Banded watersnakes (*Nerodia fasciata*) occur throughout the Coastal Plain anywhere there is freshwater habitat. The eyes of this snake look cloudy because it is molting its skin, including even the protective membrane covering its eyes. Although banded watersnakes are not venomous, humans often confuse them with cottonmouths and kill them by mistake.

Logging hardwoods in Altamaha wetlands not only fragments the forest and destroys habitat, it also impairs water quality as machinery churns the earth into a mudhole and disrupts the natural flow of water between the wetlands and the river.

The red-headed woodpecker (*Melanerpes erythrocephalus*) is one of the most easily recognized birds in Georgia, with its red head and bold black-and-white plumage. A cavity nester, it flourishes in a wide variety of habitats, including flooded woodlands. Here the red-headed woodpecker is inspecting a nest cavity currently occupied by young red-bellied woodpeckers (*Melanerpes carolinus*).

Found in wet pinelands, savannas, and bogs, the pine lily (*Lilium catesbaei*) accents the Altamaha landscape in late summer.

Approaching the size of a bullfrog, the river frog (*Rana heckscheri*) is typically found in oxbow lakes and languid backwaters. The adult puts up little resistance to capture, but its tadpoles are another story. Hundreds of huge tadpoles, each up to four inches long, congregate into a school whose mass may cause predators to perceive them not as individuals, but as something far too large to tackle. If startled, the school bursts and tadpoles streak in all directions, which may confuse a predator long enough for them to make good their escape.

The aptly named bullfrog (*Rana catesbiana*) is America's largest frog and is commonly found throughout the Altamaha basin and the rest of Georgia. The male frog advertises himself to females with a resonant "*jug-o-rum*" call.

The yellow-bellied slider (*Trachemys scripta*) is one of the Altamaha's most common turtle species. This female has left the river for higher ground where she has dug a burrow to accommodate six to ten eggs. Although her hatchlings will be near-total carnivores, adult sliders are herbivores, eating early in the morning and spending much of the day basking in the sun. *Placobdella* leeches are the brown globes attached to her shell.

Perhaps the most beautiful duck found on the Altamaha, the wood duck (*Aix sponsa*) feeds not only on aquatic plants but also combs the woods for acorns, beechnuts, and wild fruit. These ducklings already have had an exciting introduction to the world. The parents nest as high as fifty feet above the ground in hollow trees or cavities excavated by pileated woodpeckers. Shortly after the ducklings hatch, the female duck calls to them from the ground until they all tumble in freefall to the forest floor.

Especially on summer weekends, the Altamaha's accessible sandbars lure people to camp out, fish, swim, and relax by the river, even within sight of the Rayonier pulp mill in Jesup. While some leave litter and cut up the shoals with four-wheeler tracks, most are good stewards of this river they call home. Just downstream, however, it's another story.

The dark color hugging the right bank of the Altamaha far downstream is not a shadow but industrial wastes discharged by Rayonier. Although legally in compliance with the terms of its permit, the mill's discharge nevertheless impairs water quality and has major negative impacts on aquatic life. Downstream from the outfall, anglers and boaters find the odor overwhelming and fish caught in this section of the Altamaha are unpalatable in the extreme.

The only purple and pink dragonfly in the southeast, the roseate skimmer (*Orthemis ferruginea*) is attracted to all size bodies of fresh water, from small puddles to lakes. The males choose sites with a prominent perch, which they aggressively defend against other males.

Native water lilies (*Nymphaea odorata*) are commonly found in still backwaters and ponds along the length of the Altamaha. This pink flower is a rare beauty among the far more common white lilies.

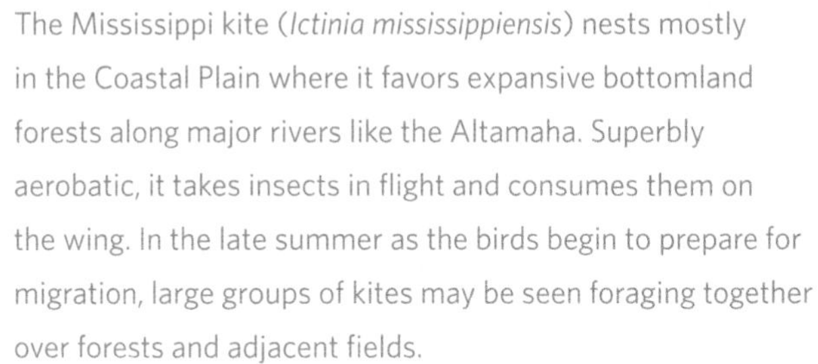

The Mississippi kite (*Ictinia mississippiensis*) nests mostly in the Coastal Plain where it favors expansive bottomland forests along major rivers like the Altamaha. Superbly aerobatic, it takes insects in flight and consumes them on the wing. In the late summer as the birds begin to prepare for migration, large groups of kites may be seen foraging together over forests and adjacent fields.

While all other brown-backed thrushes migrate to Central and South America for the winter, the hermit thrush (*Catharus guttatus*) winters in the Altamaha bottomlands. Although it has a melodious ringing call during the breeding season, in winter the bird is very quiet and secretive as it hunts for insects on the forest floor.

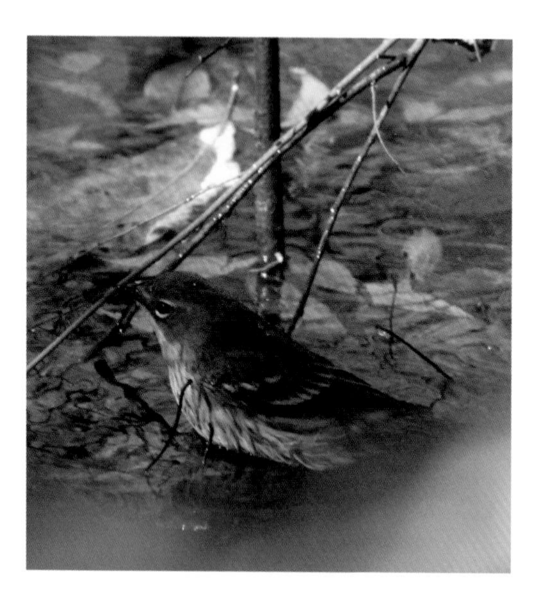

The yellow-rumped warbler (*Dendroica coronata*) is a common winter resident across the state. Here the bird is bathing in a shallow pool warmed by the winter sun.

This pair of summer tanagers (*Piranga rubra*) illustrates the striking difference between the plumage of the male and female birds; not surprisingly, it is the well-camouflaged female who incubates the eggs. The male bird is an insistent singer whose "*picky-tuck*" is heard throughout the summer even if he remains well concealed. Early in the breeding season, the birds feed extensively on bees and wasps, balancing that diet with more fruit as it becomes available in summer.

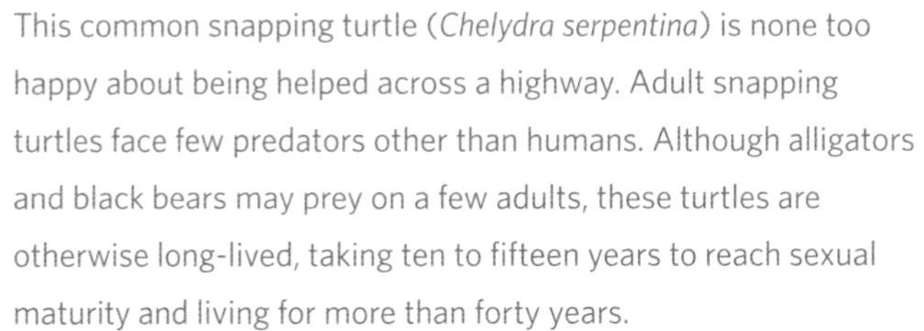

This common snapping turtle (*Chelydra serpentina*) is none too happy about being helped across a highway. Adult snapping turtles face few predators other than humans. Although alligators and black bears may prey on a few adults, these turtles are otherwise long-lived, taking ten to fifteen years to reach sexual maturity and living for more than forty years.

A winter resident of the Altamaha, the ruby-crowned kinglet (*Regulus calendula*) gleans insects and spiders from tree branches, leaves, and needles, even taking advantage of available fruit and tree sap. Although one of the smallest birds found along the river, it makes its presence known by its active foraging, flashing its wings to flush prey from hiding.

The pineland hibiscus (*Hibiscus aculeatus*) raises its flowers high above other plants in the Altamaha wetlands. Its flowers last only a day and then twist closed like a paper bag.

The anhinga (*Anhinga anhinga*) is commonly known as the "snake bird" since it is usually seen swimming with only its slender neck and head visible above water. It often shares rookery trees with herons and wood storks. This anhinga in showy nuptial plumage (*above*) must perch to allow its waterlogged feathers to dry before it can take flight.

During the mating season, the bellows and purrs of male alligators resound through the Altamaha bottomlands. After mating, female alligators take great care in building nests for their eggs. Hatchlings spend up to two years near their mother for protection from predators, and then they graduate to an additional year in a small group or pod before striking out on their own.

The Ohoopee sand dunes are a desert alongside a giant oasis known as the Altamaha River system. Other mysterious sand ridges cutting the river's path are the relics of ancient shorelines.—JAMES HOLLAND

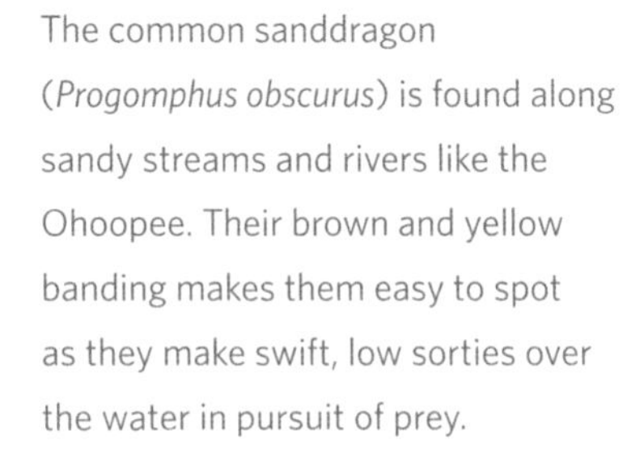

The common sanddragon (*Progomphus obscurus*) is found along sandy streams and rivers like the Ohoopee. Their brown and yellow banding makes them easy to spot as they make swift, low sorties over the water in pursuit of prey.

In the Ohoopee dunes, an assortment of reindeer lichens (*Cladina* species) carpet the well-drained, sandy soil beneath turkey oaks.

The slimleaf pawpaw (*Asimina angustifolia*) is a dwarf shrub that produces an outsized magnolia-like flower with a purple center. It favors dry, sandy woodland habitats along the Altamaha. Birds and small mammals feed on its oblong, yellow-green fruit.

The Ohoopee River is referred to as a "blackwater" river because its waters are stained brown with tannic acids released from decaying vegetation in the heavily forested areas it drains. Here at the confluence, its dark waters blend quickly with those of the Altamaha. An unusual feature lies along the Ohoopee's eastern shore: twenty-two thousand acres of sand hills deposited by strong westerly winds that blew across exposed river sands at the height of the last ice age approximately twenty thousand years ago. Fast-draining, nutrient-poor soils of the Ohoopee dunes stunt the growth of trees and shrubs, but rare plants that can cope in this humid desert abound.

The gopher tortoise (*Gopherus polyphemus*) is widely distributed south of the Fall Line where it can dig easily into the deep, sandy soil. Its spacious burrows, averaging fifteen feet long and penetrating six feet below ground, provide the tortoise with far more comfortable temperatures and humidity than it experiences at the surface. As many as three hundred species use the burrow as well, and some, like the indigo snake, completely depend on the gopher tortoise to provide them shelter. Because so many other animals rely on these burrows, the gopher tortoise is often referred to as a keystone species.

A male hatchling takes sixteen to eighteen years to reach sexual maturity, a female as long as twenty-one years. Gopher tortoise numbers are in decline due to a combination of natural factors, such as high nest predation, slow growth rate, and habitat destruction.

The southeastern spinyleg dragonfly (*Dromogomphus armatus*) breeds along small seeps or streams. Butterflies are a favorite prey of this rare dragonfly that occurs only in scattered locations below the Fall Line, such as here in the dunes bordering the Ohoopee River near its confluence with the Altamaha.

With iridescent eyespots adorning its wings, the common buckeye butterfly (*Junonia coenia*) is far more stunning than its name would suggest. A butterfly usually found in open areas, the males often perch for long periods to watch for females. Here the butterfly is feeding on clusterleaf blazing star (*Liatris laevigata*), which flourishes on exceptionally well-drained soils.

In ancient Egypt, the scarab beetle rolling its ball of dung was seen as an earthly manifestation of the movement of the sun across the heavens. Along the Altamaha, scarab beetles are less revered. Known as "tumble-bugs," these beetles gather a ball of fresh manure, roll it some distance, dig a shallow hole to accommodate the ball, and lay an egg inside it just before backfilling the hole. Not only do dung beetles protect the next generation through these actions, they also aerate the soil and recycle nutrients.

Sometimes you need an entomologist along to know exactly which insect you have found in the Altamaha basin. That failing, this stout cricket can be identified in more general terms only as a member of the order Orthoptera, family Gryllidae.

Found on two small plots of private land in the sand-ridge region and nowhere else in the world, Radford's mint (*Dicerandra radfordiana*) is the rarest of the Altamaha's endemic plants. Its showy fuchsia blossoms attract pollinators like this gulf fritillary (*Agraulis vanillae*), essential to producing the plant's pinhead-sized seeds. Each year, scientists from the Georgia Department of Natural Resources join with volunteers to conduct a census of each tract to establish population numbers that vary widely, depending on summer rainfall.

These are the feathery seedpods of the vine *Clematis reticulata*, commonly known as netleaf leatherflower, which favors dry woods and sandy soils along the Altamaha. Lacking tendrils, its leafstalks bend and clasp, enabling the vine to climb toward the sun.

Although known commonly as blue-eyed grass, *Sisyrinchium albidum* is a member of the lily family and not a grass at all. It prefers open, dry woods with sandy soil.

Near Townsend, the Altamaha River watershed is home to the largest old-growth cypress in the eastern United States. Some of these trees are believed to be more than one thousand years old. When visiting these ancient reminders of what was, you will stand in awe at the majesty of their size, beauty, and gracefulness.—JAMES HOLLAND

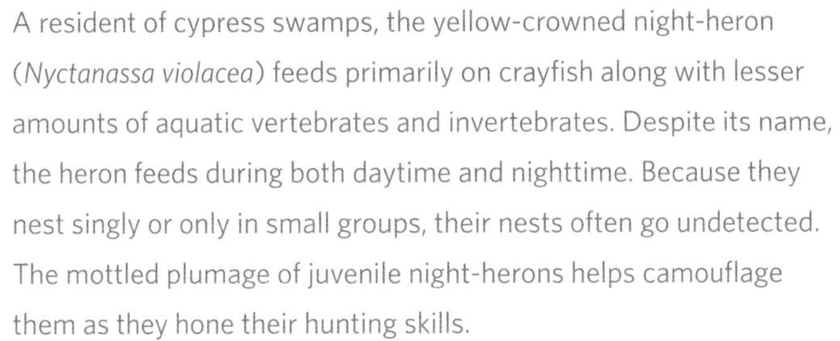

A resident of cypress swamps, the yellow-crowned night-heron (*Nyctanassa violacea*) feeds primarily on crayfish along with lesser amounts of aquatic vertebrates and invertebrates. Despite its name, the heron feeds during both daytime and nighttime. Because they nest singly or only in small groups, their nests often go undetected. The mottled plumage of juvenile night-herons helps camouflage them as they hone their hunting skills.

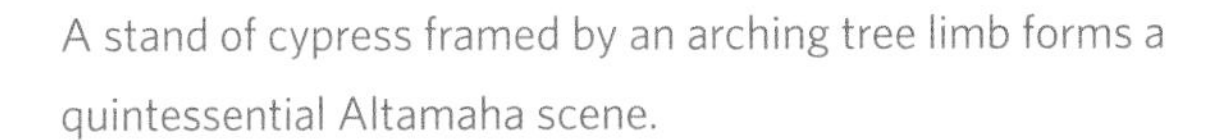

A stand of cypress framed by an arching tree limb forms a quintessential Altamaha scene.

With the Altamaha in flood, the bottomlands become a sheet of moving water surging among trunks of bald cypress and Ogeechee tupelo.

Bald cypress are well adapted to withstand as much as a six-foot fluctuation in water levels, depending on the season.

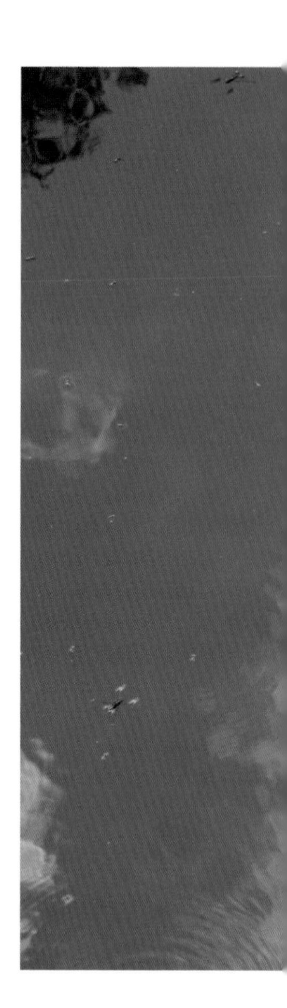

In the Altamaha back swamps of Long County, a stand of giant bald cypress (*Taxodium distichum*) includes the largest specimen in Georgia, with a girth of forty-four feet, five inches. In 2010, this tract was purchased from private owners and incorporated into the Townsend Wildlife Management Area. Sunset burnishing the treetops reflects in the dark waters below.

Sometimes the water surrounding Georgia's largest cypress is so deep that visitors can approach only by boat. But even waist-deep water fails to stop some people from visiting the tree.

Another giant cypress easily shelters three adults within its tripod trunk. The hollow may be the result of a lightning strike that injured but failed to kill the tree.

Surprised at its roost inside a hollow bald cypress, the Rafinesque's big-eared bat (*Corynorhinus rafinesquii*) may live up to ten years in the wild. Unlike bats that begin to emerge at dusk, this moth specialist waits until it is completely dark to hunt for insects. At rest or during hibernation, it curls its enormous ears against its head to conserve body moisture.

Some of the bald cypress trees have hollows large enough to accommodate a person inside.

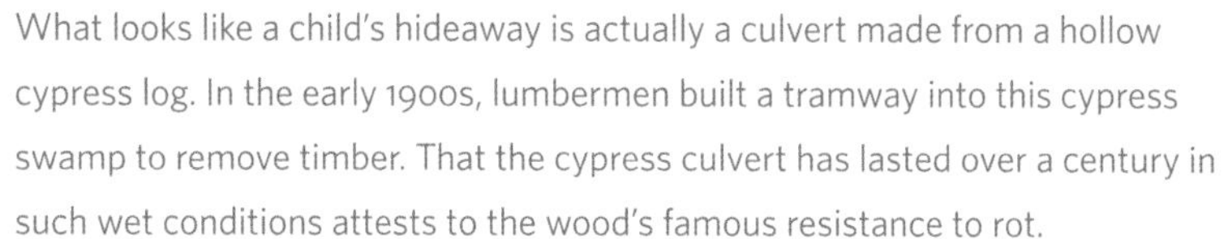

What looks like a child's hideaway is actually a culvert made from a hollow cypress log. In the early 1900s, lumbermen built a tramway into this cypress swamp to remove timber. That the cypress culvert has lasted over a century in such wet conditions attests to the wood's famous resistance to rot.

This cypress tree may have been felled and partially sawed into lengths intended for use as culverts.

In the early 1900s when this cypress tract was cut for timber, sawyers abandoned their efforts to fell this particular tree, perhaps because it was too large to extract from the swamp. The cypress survived despite the depth of the cuts.

American alligators provide structure to rivers and streams by excavating holes and caves where they rest and wait in ambush for prey. During times of drought, the "gator holes" can serve as indispensable watering holes for many creatures. Here an alligator lounges in a self-made mud bath under a fallen cypress during a winter dry spell.

Land cleared for new agricultural fields has left a cypress pond completely isolated and its waters fouled with muddy runoff.

Still, black water reflects the sky and buoys a floating tupelo leaf.

When you reach the tidal range, the Altamaha becomes a totally different river. Everything changes, even the fish species. The freshwater marshes mark the beginning of a most unusual change in ecosystems and are a gift that only nature could provide.—JAMES HOLLAND

As the waters of the Altamaha begin to meet the ocean, the river widens and slows its pace to form an estuary protected from the open sea by a chain of barrier islands lying offshore.

The Altamaha estuary serves as a large filter for the silt and clay brought from far inland, some of which settles out to form marsh mud. This rich sediment supports the growth of a high diversity of plant species.

The big bluet (*Enallagma durum*) is an uncommon damselfly found along brackish sections of the lower Altamaha. From a riverside perch, the male guards the female as she lays eggs in the water.

Black-tipped wings and a long, downcurved bill are hallmarks of the white ibis (*Eudocimus albus*). This ibis prefers to nest deep in Altamaha swamps but flies long distances to feed in freshwater ponds and marshes.

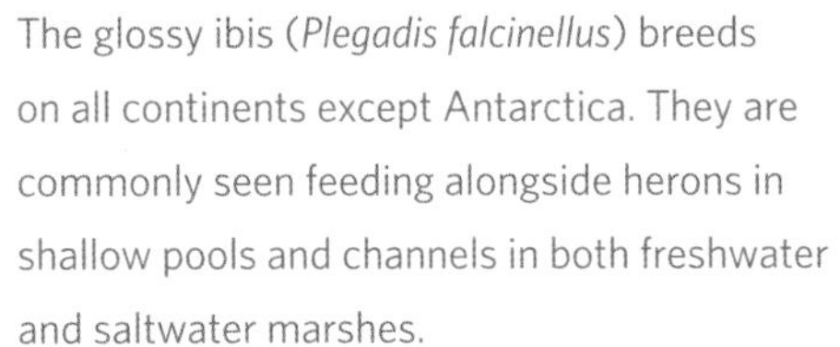

The glossy ibis (*Plegadis falcinellus*) breeds on all continents except Antarctica. They are commonly seen feeding alongside herons in shallow pools and channels in both freshwater and saltwater marshes.

This low-climbing red morning glory (*Ipomoea hederifolia*) prefers a shady, damp setting.

The bright-red, tubular flowers of the coral bean (*Erythrina herbacea*) are very attractive to hummingbirds. However, the equally eye-catching beans and leaves contain toxins that are poisonous to humans.

In an Altamaha freshwater marsh, a praying mantis awaits its next meal on a flower spike of pickerel weed (*Pontederia cordata*).

The blue butterwort (*Pinguicula caerulea*) is a wetland plant belonging to the bladderwort family, which contains many carnivorous species. Its funnel-shaped flower ensures that pollinating insects must crawl past its reproductive parts to reach nectar deep in the end of the flower.

Despite its name, this summer azure butterfly (*Celastrina neglecta*) also flies in the spring. Its caterpillars feed on dogwood leaves.

Observing the grace and agility of the swallow-tailed kite (*Elanoides forficatus*) is an unforgettable experience. They nest in the tops of the tallest trees in "pine islands" that dot the lower floodplains and swamp forests of the Altamaha, where they enjoy their greatest nesting success in Georgia. Like the Mississippi kite, they capture insects they eat on the wing and also seize snakes, frogs, and lizards from the tree canopy. These birds are very gregarious and tend to return year after year to the same nesting area as long as it remains undisturbed. This juvenile kite (*left*) lacks the signature outer tail feathers of an adult.

By the start of the breeding season, both male and female great egrets (*Ardea alba*) sport a nuptial train of up to fifty-four long, filamentous plumes. Egrets display their magnificent nuptial plumage both to attract a mate and to reinforce the pair bond during the mating season. In addition to feeding in shallow water, great egrets often are seen feeding in the uplands. At the turn of the twentieth century, great egrets were killed by the thousands during breeding season so that their filmy feathers could be used to adorn ladies' hats. The slaughter was so great that it stimulated the creation of a national conservation effort to protect the birds, both through federal law and by means of the work of conservation organizations. The National Audubon Society uses the great egret as its symbol.

The sunrise-pink feathers and grey-green banjo-shaped beak make it impossible to mistake the roseate spoonbill (*Ajaia ajaja*) for any other bird. It swings its partially open bill from side to side to feel for fish and crustaceans, and on contact with prey its bill snaps shut. Although decimated by plume hunters in the early twentieth century, with protection spoonbills have rebounded spectacularly, expanding their range from Florida into Georgia.

The wood stork (*Mycteria americana*) is the only North American representative of the world's stork family. The naked skin on its head has led to its common name "flinthead." Wood storks ride rising columns of warm air to soar efficiently back and forth to their feeding grounds. When feeding, wood storks wade through shallow water with their bills submerged, waiting to detect the movement of small fish and frogs. As soon as it contacts prey, the bill reflexively snaps shut in twenty-five thousandths of a second, one of the fastest reflexes known in the vertebrate world.

Wood storks are highly gregarious birds, often nesting not only with other wood storks but with other wading birds. Wood stork pairs often mate for life and return to the same nest each year, which they refurbish with sticks and other nesting material.

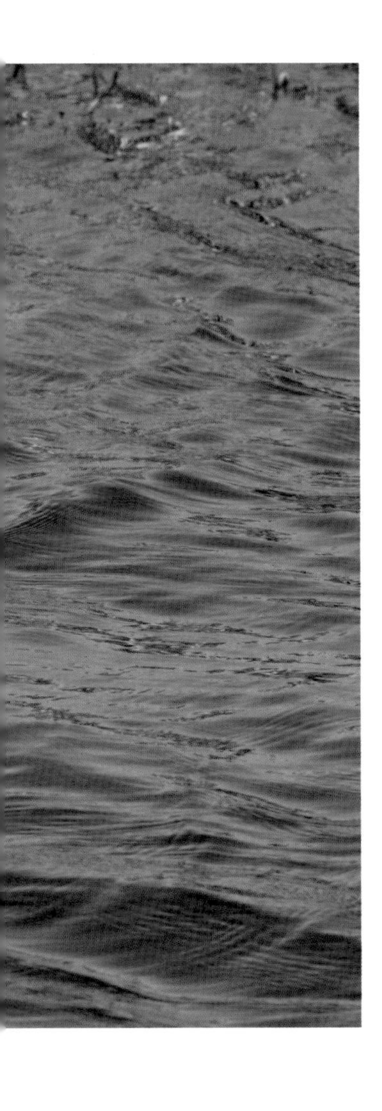

Wading birds like herons and egrets favor a rookery tree surrounded by water patrolled by alligators, although the birds run the risk that some nestlings will fall into the water and be preyed upon by the reptiles. On balance, the birds benefit because the presence of alligators deters aggressive nest predators like raccoons (*Procyon lotor*) who readily take to water.

Fungi like these golden, trumpet-shaped mushrooms and pompom mushrooms help to break down plant matter alongside quiet blackwater streams entering the Altamaha west of Darien.

These rare Bartram's airplants (*Tillandsia bartramii*) are epiphytes, plants that use other plants or trees for support structures but do not parasitize them. Like their famous relative Spanish moss, airplants are members of the bromeliad family and closely related to pineapples.

The long-tailed skipper butterfly (*Urbanus proteus*) is found commonly in open, brushy areas where it feeds on nectar from wildflowers like this coreopsis. Its caterpillars relish vining legumes like peas and beans where they can become a pest. They do have the redeeming quality of also eating non-native wisteria.

Introduced by humans, non-native plants can wreak havoc with the Altamaha ecosystem in the absence of pests and diseases that may have kept them in check in their original home. Once they escape from cultivation, non-native plants such as kudzu, Chinese privet, Japanese honeysuckle, Chinese wisteria, and tallow trees smother, strangle, shade, and crowd out native plants and trees. Pollinators like this silver-spotted skipper (*Epargyreus clarus*) feeding on Chinese wisteria (*Wisteria chinensis*), as well as animals that eat fruit and disperse seeds, unwittingly assist in broadening the range of these noxious plants. In contrast, some non-native plants such as this Indian turnsole (*Heliotropium indicum*) are not rampant colonizers and thus have limited distribution with no known harmful effects.

Dug as straight as a rifle shot, Rifle Cut was supposed to redirect the flow of the Altamaha to bypass an inconvenient sand shoal just downriver from its confluence with the Couper River. Because the soil was matted thick with cypress roots, hand labor could not excavate deep enough to divert the Altamaha from its natural course. Although plans to deepen the cut continued into the 1880s, they never got off the drawing board due to engineering difficulties and competition from the railroads.

This unnaturally straight path through the cypress is all that remains of the Brunswick-Altamaha Canal. Plans for cutting a canal date as far back as 1798, when traders realized the natural harbor at Brunswick was better than the one at Darien. Excavation finally started in October 1836 only to be thwarted by the financial Panic of 1837. For three years work proceeded intermittently until it was stopped altogether in November 1839. A new group of investors resumed digging in 1852 and opened the canal in June 1854. Because sections of the canal followed naturally sinuous tidal creeks, it was hardly a time-saver and was abandoned in 1860.

Beautifully reflected against the dark water of Rifle Cut, spider lilies (*Hymenocallis floridana*) brighten the dark shade of the cypress.

West of Darien, abandoned rice field dikes are all that remain of the rice culture that thrived from the mid-1700s until after the Civil War. Using slave labor, planters converted more than thirty thousand acres of freshwater marshes into rice plantations along the Altamaha. The flooded fields now support a wide variety of waterfowl as seen here at the Altamaha Waterfowl Management Area.

The pied-billed grebe (*Podilymbus podiceps*) is a small, compact aquatic bird seen usually in the fresh- to brackish-water sections of the Altamaha. It is an excellent swimmer, diving to feed on small fish, crustaceans, and aquatic insects. It also consumes its own feathers, perhaps as an aid to digestion. When disturbed, the grebe may dive rapidly or sink slowly out of sight and remain submerged for a long interval of time with only its beak and eyes above water.

The bobolink (*Dolichonyx oryzivorus*) visits the coastal region of the Altamaha during the spring and fall migration between its wintering grounds in South America and its breeding grounds in the northern tier of the United States and Canada, a yearly round-trip of more than twelve thousand miles. Known in tidewater Georgia as the "ricebird," the bobolink was a scourge of the rice fields, eating newly sprouted plants in the spring and feasting on the ripe grain in the fall. Planters did not hesitate to shoot them by the hundreds and many nineteenth-century cookbooks contained recipes for ricebird dishes.

A small, stout member of the rail family, the sora (*Porzana carolina*) winters among the dense shoots of marsh grasses where its mottled plumage is excellent camouflage. It prefers to walk rather than fly, flicking its short tail to flash white feathers as it searches for the seeds that make up the bulk of its diet.

An egret, a heron, and turtles bask in the warmth of the February sun.

Paper wasps (*Polistes* species) construct a home for the next generation under a palmetto frond.

Protecting freshwater wetlands helps protect rare butterflies, too. The palmetto skipper (*Euphyes arpa*) is feeding on one of its favorite nectar sources, pickerel weed flowers, a characteristic wetland plant. Palmetto skipper caterpillars eat saw palmetto fronds, hence its common name.

The seaside sparrow (*Ammodramus maritimus*) is most commonly seen in brackish and saltwater marshes, where it nests in shorter vegetation above the reach of the highest tides. Omnivores, seaside sparrows take advantage of the presence of insects and crabs in the marsh as well as seeds of various marsh plants. In addition to threats from tropical storms and hurricanes, these birds are vulnerable to degradation of marsh habitat caused by a wide array of human activities.

The American bittern (*Botaurus lentiginosus*) is rarely sighted among the cattails and bulrushes of the Altamaha's freshwater marshes, where it spends the winter. It is a stealth predator, standing motionless and eyeing its prey until striking with a rapid thrust of its bill. Bitterns feed mostly on frogs and small fish.

Constructed in 1721 near what would become the site of Darien, the garrison at Fort King George served for seven years on the frontier of the British Empire in America. Built of cypress cut along the Altamaha, the fort was located here to ward off French expansion into the region and to serve as an early warning system for potential incursions by the Spanish. In 1988 the blockhouse and fortifications were carefully reconstructed and the site is permanently protected as a state park.

As I observe this vast sea of beautiful marshes, a calmness comes over my body like a gentle, cool breeze. The salt marshes and tidal creeks have supported me and my family for more than three decades. The health of the salt marsh depends on the quality of water coming in.—JAMES HOLLAND

Shrimpers operating out of Darien depend on the coastal marshes to serve as nurseries for the shrimp that ultimately fill their nets.

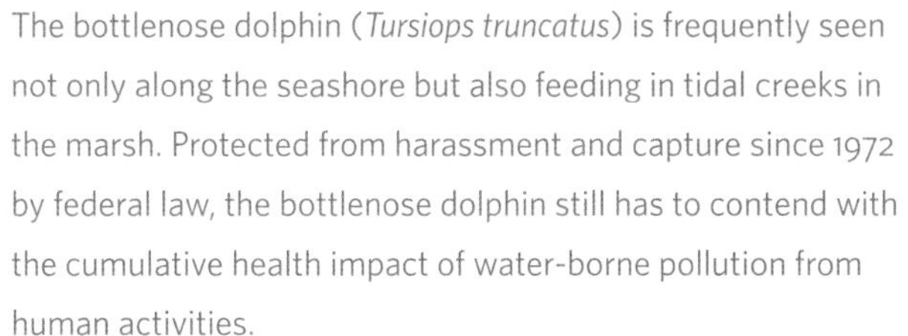

The bottlenose dolphin (*Tursiops truncatus*) is frequently seen not only along the seashore but also feeding in tidal creeks in the marsh. Protected from harassment and capture since 1972 by federal law, the bottlenose dolphin still has to contend with the cumulative health impact of water-borne pollution from human activities.

An animal up to fifteen feet long and weighing over three thousand pounds might seem hard to miss, but the West Indian manatee (*Trichechus manatus*) hides most of its bulk underwater. Here only the noses of a mother manatee and her calf show above water as they join a third manatee to browse along the edge of a tidal creek.

A flock of American avocets flies past the Sidney Lanier Bridge before settling into the marsh.

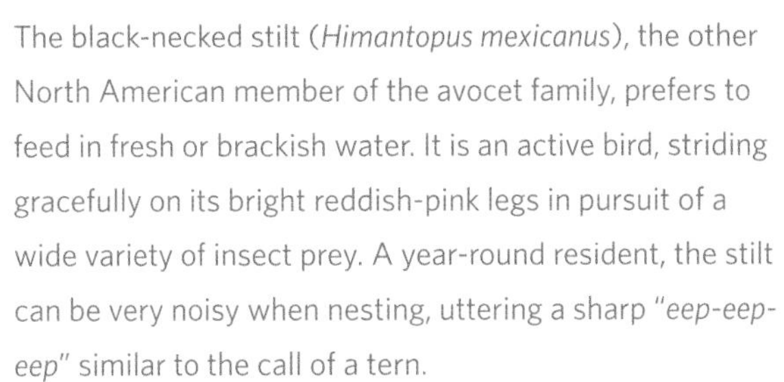

The black-necked stilt (*Himantopus mexicanus*), the other North American member of the avocet family, prefers to feed in fresh or brackish water. It is an active bird, striding gracefully on its bright reddish-pink legs in pursuit of a wide variety of insect prey. A year-round resident, the stilt can be very noisy when nesting, uttering a sharp "*eep-eep-eep*" similar to the call of a tern.

Despite its delicate appearance, the American avocet (*Recurvirostra americana*) migrates thousands of miles from the Great Plains and intermontane West to winter in Georgia's marshes. The bird sweeps its upturned bill from side to side in shallow water to stir up small crustaceans and insects as well as the water plants on which it feeds. Avocets are very social, often feeding in flocks of hundreds standing shoulder to shoulder as if moving in formation.

Fiddler crabs swarm marshes by the billions, feeding on detritus they extract from marsh sediment. The males gesture with their enlarged claw as a signal to attract females and to ward off rival males.

With flashing red and yellow epaulets and a cheery *"o-ka-lee,"* the male redwing blackbird (*Agelaius phoeniceus*) heralds the arrival of spring along the Altamaha.

Putting the lie to the assumption that owls are supposed to nest in hollow trees, a pair of great horned owls (*Bubo virginianus*) commandeered an osprey nest atop a marsh-side platform to raise their young. The great horned owl eats mostly small mammals but will prey on birds, reptiles, amphibians, and even invertebrates. This nestling should begin to fly at seven weeks of age and may remain with its parents for the rest of the summer.

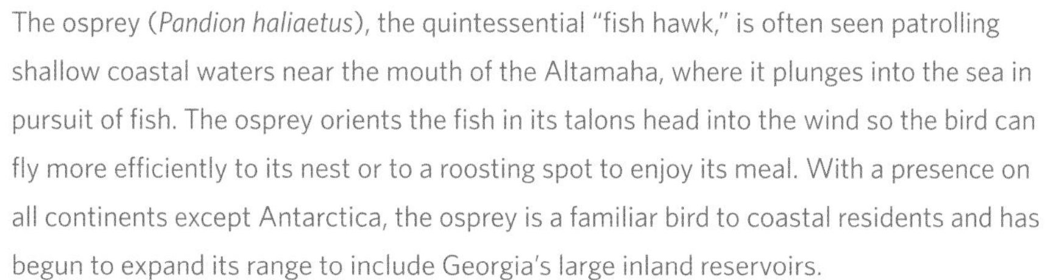

The osprey (*Pandion haliaetus*), the quintessential "fish hawk," is often seen patrolling shallow coastal waters near the mouth of the Altamaha, where it plunges into the sea in pursuit of fish. The osprey orients the fish in its talons head into the wind so the bird can fly more efficiently to its nest or to a roosting spot to enjoy its meal. With a presence on all continents except Antarctica, the osprey is a familiar bird to coastal residents and has begun to expand its range to include Georgia's large inland reservoirs.

Although the tree swallow (*Tachycineta bicolor*) nests only north of the Fall Line in Georgia, it is a prominent winter resident of the Altamaha salt marshes, where great clouds of the birds swoop over the marshes eating insects on the fly. In addition to live prey, tree swallows descend in roiling masses onto wax myrtle bushes where they gorge on spicy berries.

The Forster's tern (*Sterna forsteri*) winters along the Georgia coast where it is often observed catching insects on the wing above the Altamaha marshes. As with most terns, however, fish make up the bulk of its diet. This bird in winter plumage shows the characteristic narrow black band across and behind the eyes.

At this site where the lower Altamaha is cutting into its bank, oyster shells cascade toward the water. These shells are not refuse from a recent oyster roast but from ancient meals, perhaps as much as five thousand years old. At that time, sea level was stabilizing following the end of the last ice age. The ocean had flooded the area behind coastal islands to form salt marsh lagoons, ideal environments for the development of extensive oyster beds. Indians already residing at the coast began to exploit this new bonanza of shellfish. Called "middens" by archaeologists, these mounds are composed mostly of discarded shells from oysters and other shellfish but also contain mammal, fish, and bird bones as well as occasional fragments of pottery, carved bone ornaments, and worked stone tools. Some individual middens are quite large; one on Sapelo Island rises six to ten feet above the land surface and is nearly the length of a football field. Other midden sites, particularly those facing the marsh, may consist of many small shell mounds scattered over a large area. The midden shown here lies ten miles upriver from the nearest oyster beds today. We know nothing of its age, only that the Altamaha will soon reclaim its relics.

Its richly iridescent plumage alone would make the male boat-tailed grackle (*Quiscalus major*) one of the Altamaha's most striking birds. Seldom at rest, these birds accent their activities with piercing whistles and creaks and interact boldly with humans in the coastal zone. Males and females band together only during the breeding season, when males attract a harem of females to form a nesting colony.

Beds of eastern oysters (*Crassostrea virginica*) lining a tidal creek in the salt marsh sit high and dry during low tide. In the early 1900s, Georgia led the nation in the commercial harvest of oysters. Today the industry is nearly nonexistent due to overharvesting, pollution, and disease.

Like many ambush hunters, the green heron (*Butorides virescens*) is seldom seen but widely distributed in the Altamaha basin. A solitary hunter, the green heron is among the few tool-using birds. It has been observed laying "bait" such as twigs, feathers, earthworms, insects, and even bread on the water surface to lure fish to within its striking distance.

The yellow rat snake (*Elaphe obsoleta quadrivittata*) is a superb climber, widely distributed in wooded habitats and suburban settings in the Altamaha basin. In addition to eating rodents, adult rat snakes eat other small mammals, birds, and bird eggs. If threatened, rat snakes freeze and kink their body lengthwise like an accordion. Herpetologists believe the snake's pose may fool a predator into thinking it is a stick, not a meal.

The Altamaha, which supports a lively recreational fishery in both its freshwater and salt-water reaches, not only benefits anglers but also brings needed revenue to local communities. James Holland's daughter, Kathy Dubose, beams with pride over her 19.5-inch-long speckled trout.

This small flock of double-crested cormorants (*Phalacrocorax auritus*) has come to roost for the evening after a long day in pursuit of fish. They are often seen flying overhead in a V or in a single-file skein. Although they spend much of their time in the water, their feathers lack waterproofing oils. At intervals cormorants must emerge from the water to dry their wings in the sun.

Because of the effects of the pesticide DDT, the bald eagle (*Haliaetus leucocephalus*) had disappeared from Georgia by the early 1970s. Unprecedented conservation efforts helped restore our national bird to Georgia skies. While bald eagles nest most commonly in the Altamaha's coastal zone, they have expanded their range upriver to include stretches of the Ocmulgee and Oconee Rivers as well as manmade reservoirs.

The great blue heron (*Ardea herodias*) adapts readily to both freshwater and saltwater habitats and can be seen widely across Georgia. Its long legs allow it to exploit deeper water, where it forages for a wide variety of fish and other organisms. Because it stands quietly to ambush its prey, it is often overlooked, despite its size.

What appears to be a double exposure is a molting blue crab in the process of casting off its old shell. The molt is a difficult time for the famously feisty crab since they are particularly vulnerable to predators until the new shell hardens.

Abandoned or lost fishing gear, such as this crab trap, will continue to "ghost fish"—trap fish or crabs that cannot escape—until the equipment finally deteriorates. This great egret now uses the abandoned gear for its own fishing forays.

The clapper rail (*Rallus longirostris*) is often heard but seldom seen. Its laterally compressed body allows it to move easily through the dense salt marsh it favors, hence the expression "thin as a rail." While it is otherwise secretive, its clattering advertisement and territorial calls punctuate the marsh soundscape.

The willet (*Tringa semipalmata*) is often observed walking along the surf line alone or in small groups searching for small crustaceans and other invertebrates. These large shorebirds also feed along tidal creeks, oyster beds, and in less-vegetated areas of the salt marsh. Wary of humans, the mottled grey willet is unmistakable when it lifts its wings to reveal a distinctive white W bordered by black as it takes a short flight to resettle farther down the beach.

The perennial salt marsh aster (*Symphyotrichum tenuifolium*) is a favorite of butterflies along the edge of the salt marsh. The plant is adapted to tolerate occasional inundation by the tide.

A new addition to the Altamaha coastal region is the white pelican (*Pelecanus erythrorhynchus*), which has begun to winter here in the hundreds. With a wingspan of up to nine feet, these pelicans are among the largest living birds, much larger than the more familiar brown pelican. Unlike the plunge-diving brown pelican, the white pelican fishes from the surface using its expansive throat pouch like a dip net. It may also cooperate with other white pelicans to surround fish and herd them close together so they have less chance of escape.

These "Salt Marsh Soldiers" represent a new generation of stewardship for the Altamaha. Vicki Klahn of Glynn Middle School leads her students into hands-on marsh ecology studies in the field and guides them in service projects, such as marsh cleanups, benefiting all coastal residents.

An autumn sun sets over the *Spartina* along a tidal creek.

The Altamaha's barrier islands are a lovely gift from nature for the migrant birds and other wildlife that use them to rest and nourish while traveling thousands of miles to reach their predetermined destinations. This wild place is wholly given over to nature.—JAMES HOLLAND

As one of four barrier islands accessible by automobile, St. Simons Island underwent unprecedented residential development during the last twenty years. In the center of the photo, the recreational beach is very narrow because of the armoring of the coastline to protect property. In the foreground is Gould's Inlet, where an enormous amount of sand has accumulated on the beach and in offshore shoals since Hurricane Dora in 1964.

Little St. Simons Island, a privately-owned, ten-thousand-acre barrier island, is an excellent example of the salt marsh–barrier island ecosystem. The salt marsh, cut by tidal creeks, gives way to the maritime forest occupying the uplands. Along the island's shore are dunefields, sand spits, and a seven-mile-long beach. With the exception of a small residential compound, this undeveloped island provides a sharp contrast to other Altamaha islands accessible by car.

Although the state must approve all plans for construction in the marsh in order to minimize environmental impact, some property owners violate the terms of their permits or illegally build without a permit. The aerial view shows the tracks left by a crane used to construct a dock. Ignoring best construction practices, the crane operator left a path of destruction extending all the way to a tidal creek. Without applying for an additional mandatory permit, the property owner also built an illegal bulkhead behind the house in the marsh buffer zone, ostensibly to protect his property from erosion, although the tidal creek is 750 feet away. At marsh level, deep ruts, crushed vegetation, and the compacted marsh mud left in the crane's wake are obvious. Repairing the damage is costly, and restoration of the marsh to its natural state is not guaranteed.

Beachfront residential development often fails to take into account how dynamic barrier islands are. Seawalls armoring the coast may protect homes but the waves deflected by seawalls destroy the recreational beach, shifting beach sand offshore.

Historically found only along the U.S.-Mexico border, the nine-banded armadillo (*Dasypus novemcinctus*) is rapidly expanding its range northward into Georgia. Armadillos now occur throughout the Altamaha basin, including its barrier islands.

The snowy egret (*Egretta thula*) is easy to distinguish from other white egrets by its black legs and bright yellow feet. A very social bird, the snowy egret feeds in shallow water, both salt and fresh, where it forages often in large mixed-species flocks.

The brown pelican (*Pelecanus occidentalis*) is one of the most familiar seabirds of the Georgia coast. Little Egg Island Bar, at the mouth of the Altamaha, supports one of the largest pelican rookeries in the Southeast, as many as twenty-eight hundred nests per year. The pelican's throat pouch indeed can hold more than its belly can—up to three gallons of water and fish. The bird allows water to drain from the corners of the mouth so that it swallows mostly fish, even very large ones, as seen here.

One of the most beautiful of the Altamaha's butterflies, the gulf fritillary (*Agraulis vanillae*) lays its eggs on passionflowers and other members of the maypop family. Here, the butterfly poses atop the seedheads of sea ox-eye daisies (*Borrichia frutescens*). Bright colors warn that toxins in its flesh make the gulf fritillary unpalatable to most potential predators.

A common winter resident, the ruddy turnstone (*Arenaria interpres*) roots through seaweed and *Spartina* stranded at the tideline in search of "sand fleas" and other amphipods. Not fussy about food choices, it also probes with its beak into the sand searching for horseshoe-crab eggs and lives up to its name by turning over small pebbles and other debris in search of a meal.

Distinguished by their plump black-and-white bodies and orange, clownlike beaks, American oystercatchers (*Haematopus palliatus*) are familiar sights on coastal beaches and marshes. They do not literally "catch" oysters; instead they use their powerful beaks to pry or pound open bivalves of all types and to probe for other food items, such as marine invertebrates. Juvenile American oystercatchers banded at the mouth of the Altamaha as nestlings have been later sighted nesting as far away as Nantucket Island.

The black skimmer (*Rhynchops niger*) is hard to confuse with any other seabird. Black and white wings power the bird forward as it drags the elongate lower half of its bill in shallow water, seeming to unzip the sea. Shrimp and even fast-swimming small fish are snapped up on contact. Although the birds will not tolerate human disturbance, these stunning members of the tern family enjoy the company of their own kind, nesting in tight formation and resting ashore barely a wingspan apart.

The railroad vine (*Ipomoea pes-caprae*) is a fast-growing pioneer plant on the shorefront, helping to stabilize sand dunes with its numerous runners. The nectar found in its showy flowers attracts a wide variety of insect pollinators before the flowers close in the afternoon.

The ghost crab (*Ocypode quadrata*) builds its burrows above the tide line in the dry sand beach or dunes. The burrows often are surrounded by marble-sized sand balls, discarded after crabs have eaten all the diatoms in each serving of sand they scrape up. More active at night, ghost crabs patrolling the beach for food in the daytime depend on their sand-colored shells for camouflage and their speedy gait to avoid predators.

Where the Altamaha finally meets the Atlantic, great forces are at work. For centuries, the Altamaha has delivered to the coast vast quantities of sand derived from hundreds of miles upstream in Georgia's interior. Equally relentless, the ocean has molded that sediment into islands we now call Sapelo, Little St. Simons, Sea Island, St. Simons, and Jekyll.

Pelican Spit, a sand shoal off the south end of Little St. Simons Island, is an important nesting and roosting spot for many shorebirds and seabirds, both resident and migratory. The larger birds are brown pelicans.

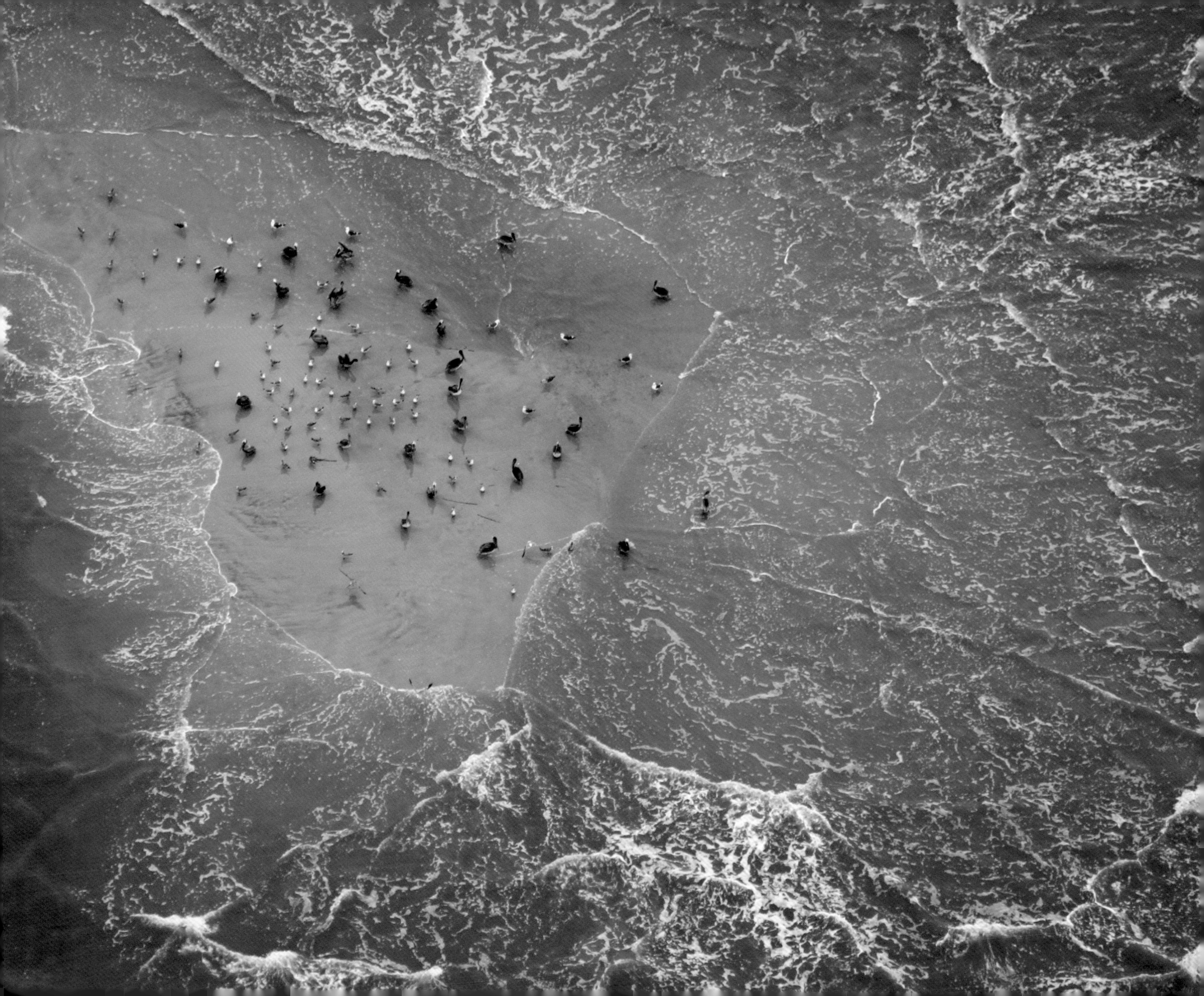

With their extensive network of leaves, stems, and roots, sea oats (*Uniola paniculata*) help to create and stabilize sand dunes that protect the natural and the built environment at the coast. As is true for many other coastal states, Georgia protects sea oats by law.

A pen shell and ark shells mark the shore where the Altamaha meets the sea.

Rare Animals and Plants of the Altamaha Basin

COMPILED BY DORINDA G. DALLMEYER

The plant and animal species listed here occur in the following twenty-eight counties lying generally south of the Fall Line in the Altamaha River basin: Appling, Baldwin, Ben Hill, Bibb, Bleckley, Candler, Coffee, Dodge, Emanuel, Glynn, Houston, Jeff Davis, Johnson, Jones, Laurens, Long, McIntosh, Montgomery, Pulaski, Tattnall, Telfair, Toombs, Treutlen, Twiggs, Washington, Wayne, Wilcox, and Wilkinson. The lists include species found in the Oconee, Ocmulgee, and Ohoopee Rivers as well as smaller tributaries.

In addition to the scientific name and the common name, the list indicates the degree to which the species is protected under federal or state law. Note that most of these plants and animals are not protected under either federal or state endangered species laws. The key to these abbreviations is as follows: Federal status: LE—listed as endangered; LT—listed as threatened; and C—candidate for listing under the Endangered Species Act. Georgia status: E—endangered; T—threatened; R—rare; and U—unusual.

Additional information on each species can be found at the Georgia Department of Natural Resources website, www.georgiawildlife.com/node/1370 (accessed July 7, 2011).

Rare Animals

GROUP	SCIENTIFIC NAME	COMMON NAME	FEDERAL STATUS	GEORGIA STATUS
Invertebrates	*Alasmidonta arcula*	Altamaha arcmussel		T
	Cambarus truncatus	Oconee burrowing crayfish		T
	Cordulegaster sayi	Say's spiketail (dragonfly)		T
	Elliptio arctata	delicate spike		E
	Elliptio congaraea	Carolina slabshell		
	Elliptio spinosa	Altamaha spinymussel	LE	E
	Fusconaia masoni	Atlantic pigtoe (mussel)		E

GROUP	SCIENTIFIC NAME	COMMON NAME	FEDERAL STATUS	GEORGIA STATUS
Invertebrates *(continued)*	*Lampsilis cariosa*	yellow lampmussel		
	Marstonia agarhecta	Ocmulgee marstonia (snail)		
	Pyganodon gibbosa	inflated floater (mussel)		
	Toxolasma pullus	Savannah lilliput (mussel)		T
Amphibians	*Ambystoma cingulatum*	frosted flatwoods salamander	LT	T
	Ambystoma tigrinum tigrinum	eastern tiger salamander		
	Desmognathus auriculatus	southern dusky salamander		
	Hemidactylium scutatum	four-toed salamander		
	Necturus punctatus	dwarf waterdog		
	Notophthalmus perstriatus	striped newt		T
	Pseudobranchus striatus striatus	broad-striped dwarf siren		
	Rana capito	gopher frog		R
	Rana virgatipes	carpenter frog		
	Stereochilus marginatus	many-lined salamander		
Fish	*Acipenser brevirostrum*	shortnose sturgeon	LE	E
	Acipenser oxyrinchus oxyrinchus	Atlantic sturgeon		
	Chologaster cornuta	swampfish		
	Cyprinella xaenura	Altamaha shiner		T
	Etheostoma parvipinne	goldstripe darter		R
	Etheostoma serrifer	sawcheek darter		
	Fundulus chrysotus	golden topminnow		
	Lucania goodei	bluefin killifish		R
	Micropterus cataractae	shoal bass		
	Moxostoma robustum	robust redhorse		E

FEDERAL STATUS LE—listed as endangered LT—listed as threatened C—candidate for listing under the Endangered Species Act
GEORGIA STATUS E—endangered T—threatened R—rare U—unusual

GROUP	SCIENTIFIC NAME	COMMON NAME	FEDERAL STATUS	GEORGIA STATUS
Fish (continued)	*Moxostoma sp. 4*	brassy jumprock		
	Notropis chalybaeus	ironcolor shiner		
	Umbra pygmaea	eastern mudminnow		
Reptiles	*Alligator mississippiensis*	American alligator		
	Caretta caretta	loggerhead sea turtle	LT	E
	Chelonia mydas	green sea turtle	LT	T
	Clemmys guttata	spotted turtle		U
	Crotalus adamanteus	eastern diamond-backed rattlesnake		
	Dermochelys coriacea	leatherback sea turtle	LE	E
	Drymarchon couperi	eastern indigo snake	LT	T
	Eumeces egregius similis	northern mole skink		
	Farancia erytrogramma erytrogramma	common rainbow snake		
	Gopherus polyphemus	gopher tortoise		T
	Heterodon simus	southern hognose snake		T
	Lepidochelys kempii	Kemp's or Atlantic ridley	LE	E
	Malaclemys terrapin	diamondback terrapin		U
	Micrurus fulvius fulvius	eastern coral snake		
	Ophisaurus attenuatus attenuatus	slender glass lizard		
	Ophisaurus compressus	island glass lizard		
	Ophisaurus mimicus	mimic glass lizard		R
	Pituophis melanoleucus mugitus	Florida pine snake		
	Rhadinaea flavilata	pine woods snake		
	Seminatrix pygaea pygaea	northern Florida swamp snake		
Birds	*Aimophila aestivalis*	Bachman's sparrow		R
	Ammodramus henslowii	Henslow's sparrow		R
	Aramus guarauna	limpkin		

GROUP	SCIENTIFIC NAME	COMMON NAME	FEDERAL STATUS	GEORGIA STATUS
Birds (continued)	*Charadrius melodus*	piping plover	LT	T
	Charadrius wilsonia	Wilson's plover		T
	Elanoides forficatus	swallow-tailed kite		R
	Falco sparverius paulus	southeastern American kestrel		R
	Haematopus palliatus	American oystercatcher		R
	Haliaeetus leucocephalus	bald eagle		T
	Himantopus mexicanus	black-necked stilt		
	Lanius ludovicianus migrans	migrant loggerhead shrike		
	Limnothlypis swainsonii	Swainson's warbler		
	Mycteria americana	wood stork	LE	E
	Nyctanassa violacea	yellow-crowned night-heron		
	Nycticorax nycticorax	black-crowned night-heron		
	Passerina ciris	painted bunting		
	Pelecanus occidentalis	brown pelican		
	Picoides borealis	red-cockaded woodpecker	LE	E
	Plegadis falcinellus	glossy ibis		
	Rynchops niger	black skimmer		R
	Sterna antillarum	least tern		R
	Sterna nilotica	gull-billed tern		T
	Tyto alba	barn owl		
	Vermivora bachmanii	Bachman's warbler	LE	
Mammals	*Corynorhinus rafinesquii*	Rafinesque's big-eared bat		R
	Eubalaena glacialis	Northern Atlantic right whale	LE	E
	Geomys pinetis	southeastern pocket gopher		T

FEDERAL STATUS LE—listed as endangered LT—listed as threatened C—candidate for listing under the Endangered Species Act
GEORGIA STATUS E—endangered T—threatened R—rare U—unusual

GROUP	SCIENTIFIC NAME	COMMON NAME	FEDERAL STATUS	GEORGIA STATUS
Mammals *(continued)*	*Lasiurus intermedius*	northern yellow bat		
	Myotis austroriparius	southeastern myotis		
	Odocoileus virginianus nigribarbis	Blackbeard's whitetailed deer		
	Trichechus manatus	West Indian manatee	LE	E

Rare Plants

GROUP	SCIENTIFIC NAME	COMMON NAME	FEDERAL STATUS	GEORGIA STATUS
Nonvascular Plants	*Brachymenium systylium*	Mexican brachymenium		
	Campylopus sp. 1	sandhill awned moss		
	Gymnocolea inflata	a liverwort		
	Teloschistes exilis	orange fructicose bark lichen		
Vascular Plants	*Acacia farnesiana*	sweet acacia		
	Aeschynomene viscidula	sticky joint-vetch		
	Agalinis aphylla	scale-leaf purple foxglove		
	Agalinis divaricata	pineland purple foxglove		
	Amorpha georgiana var. georgiana	Georgia indigo-bush	C	
	Andropogon brachystachyus	shortspike bluestem		
	Andropogon mohrii	bog bluestem		
	Armoracia lacustris	lake-cress		
	Asclepias pedicellata	savanna milkweed		
	Asimina reticulata	netleaf pawpaw		
	Asplenium heteroresiliens	Marl Spleenwort		
	Astragalus michauxii	sandhill milk-vetch		
	Balduina atropurpurea	purple honeycomb head	C	R

GROUP	SCIENTIFIC NAME	COMMON NAME	FEDERAL STATUS	GEORGIA STATUS
Vascular Plants *(continued)*	*Baptisia arachnifera*	hairy rattleweed		R
	Bouteloua curtipendula	side-oats grama		R
	Calamintha ashei	Ohoopee wild basil		R
	Callirhoe triangulata	clustered poppy-mallow	LE	E
	Calopogon multiflorus	many-flowered grass-pink		
	Carex dasycarpa	velvet sedge		
	Carex decomposita	cypress-knee sedge		
	Carex lupuliformis	hop sedge		
	Carex reniformis	reniform sedge		R
	Cayaponia quinqueloba	cayaponia		
	Ceratiola ericoides	sandhill rosemary		
	Cirsium virginianum	Virginia thistle		
	Coreopsis integrifolia	floodplain tickseed	C	T
	Crataegus triflora	three-flowered hawthorn		T
	Cuscuta harperi	Harper's dodder		T
	Cypripedium kentuckiense	Kentucky ladyslipper		T
	Dalea feayi	Feay pink-tassels		T
	Dicerandra radfordiana	Radford's mint		E
	Draba cuneifolia	wedge-leaf whitlow-grass		E
	Eleocharis albida	white spikerush		E
	Eleocharis montevidensis	spikerush		
	Elliottia racemosa	Georgia plume		
	Epidendrum magnoliae	greenfly orchid		U
	Eustachys floridana	Florida finger grass		U

FEDERAL STATUS LE—listed as endangered LT—listed as threatened C—candidate for listing under the Endangered Species Act
GEORGIA STATUS E—endangered T—threatened R—rare U—unusual

GROUP	SCIENTIFIC NAME	COMMON NAME	FEDERAL STATUS	GEORGIA STATUS
Vascular Plants (continued)	*Evolvulus sericeus var. sericeus*	creeping morning glory		U
	Forestiera segregata	Florida wild privet		U
	Fothergilla gardenii	dwarf witch-alder		
	Franklinia alatamaha	Franklin tree		T
	Galium virgatum	southwestern bedstraw		T
	Glandularia bipinnatifida var. bipinnatifida	Dakota vervain		
	Glyceria septentrionalis	floating manna-grass		
	Habenaria quinqueseta var. quinqueseta	Michaux orchid		
	Hexastylis shuttleworthii var. harperi	Harper wild ginger		
	Hibiscus grandiflorus	swamp hibiscus		
	Hottonia inflata	featherfoil		
	Hypericum denticulatum var. denticulatum	St. Johnswort		
	Hypericum erythraeae	Georgia St. Johnswort		
	Ilex amelanchier	serviceberry holly		
	Ipomoea macrorhiza	large-stem morning glory		
	Iris tridentata	savanna iris		
	Isoetes boomii	boom quillwort		
	Isoetes melanopoda	black-footed quillwort		
	Lachnocaulon beyrichianum	southern bog-button		
	Lechea deckertii	Deckert pinweed		
	Leitneria floridana	corkwood		
	Liatris pauciflora	few-flower gay-feather		
	Lindera subcoriacea	bog spicebush	C	
	Litsea aestivalis	pond spice		
	Lobelia boykinii	Boykin lobelia	C	R
	Macranthera flammea	hummingbird flower		R

GROUP	SCIENTIFIC NAME	COMMON NAME	FEDERAL STATUS	GEORGIA STATUS
Vascular Plants (continued)	*Magnolia pyramidata*	pyramid magnolia		T
	Marshallia ramosa	pineland Barbara buttons		R
	Matelea alabamensis	Alabama milkvine		R
	Matelea pubiflora	trailing milkvine		R
	Nestronia umbellula	Indian olive		R
	Oldenlandia boscii	bluets		R
	Ophioglossum engelmannii	limestone adder-tongue fern		R
	Oxypolis ternata	savanna cowbane		R
	Palafoxia integrifolia	palafoxia		
	Panax quinquefolius	American ginseng		
	Peltandra sagittifolia	arrow arum		
	Penstemon dissectus	cutleaf beardtongue		R
	Phaseolus polystachios var. sinuatus	trailing bean-vine		R
	Physostegia leptophylla	narrowleaf obedient plant		R
	Piloblephis rigida	pennyroyal		R
	Pilularia americana	American pillwort		
	Plantago sparsiflora	pineland plantain		
	Platanthera integra	yellow fringeless orchid		
	Platanthera nivea	snowy orchid		
	Polygala balduinii	white milkwort		
	Polygala leptostachys	Georgia milkwort		
	Polygonum glaucum	sea-beach knotweed		
	Ponthieva racemosa	shadow-witch orchid		
	Portulaca biloba	grit portulaca		

FEDERAL STATUS LE—listed as endangered LT—listed as threatened C—candidate for listing under the Endangered Species Act
GEORGIA STATUS E—endangered T—threatened R—rare U—unusual

GROUP	SCIENTIFIC NAME	COMMON NAME	FEDERAL STATUS	GEORGIA STATUS
Vascular Plants *(continued)*	*Portulaca umbraticola ssp. coronata*	wingpod purslane		
	Psilotum nudum	whisk fern		
	Pteroglossaspis ecristata	crestless plume orchid		
	Quercus austrina	bluff white oak		
	Quercus chapmanii	Chapman oak		
	Quercus similis	swamp post oak		
	Quercus sinuata	Durand oak		
	Rhexia aristosa	awned meadowbeauty		
	Rhexia nuttallii	Nuttall meadowbeauty		
	Rhynchospora crinipes	bearded beaksedge	C	
	Rhynchospora culixa	Georgia beaksedge		
	Rhynchospora decurrens	swamp-forest beaksedge		
	Rhynchospora macra	southern white beaksedge		
	Rhynchospora punctata	pineland beaksedge		
	Rivina humilis	rouge plant		
	Ruellia noctiflora	night-blooming wild petunia		
	Sageretia minutiflora	climbing buckthorn		
	Salix floridana	Florida willow	C	
	Sapindus marginatus	soapberry		T
	Sarracenia flava	yellow flytrap		U
	Sarracenia minor var. minor	hooded pitcherplant		U
	Sarracenia psittacina	parrot pitcherplant		T
	Sarracenia rubra	sweet pitcherplant		T
	Schisandra glabra	bay star-vine		T
	Schoenolirion albiflorum	white sunnybell		T
	Scutellaria altamaha	Altamaha skullcap		T

GROUP	SCIENTIFIC NAME	COMMON NAME	FEDERAL STATUS	GEORGIA STATUS
Vascular Plants *(continued)*	*Scutellaria drummondii*	Drummond's skullcap		T
	Scutellaria mellichampii	Mellichamp's skullcap		T
	Scutellaria ocmulgee	Ocmulgee skullcap	C	
	Sida elliottii	Elliott's fanpetals		T
	Sideroxylon macrocarpum	Ohoopee bumelia		T
	Sideroxylon thornei	swamp buckthorn	C	R
	Silene ovata	ovate catchfly		R
	Silene polypetala	fringed campion		R
	Smilax lasioneuron	carrion-flower		R
	Spermacoce glabra	smooth buttonweed		R
	Spermolepis inermis	rough-fruited spermolepis	LE	E
	Sporobolus pinetorum	pineland dropseed		
	Sporobolus teretifolius	wire-leaf dropseed	C	
	Stewartia malacodendron	silky camellia		
	Stokesia laevis	Stokes' aster		
	Stylisma pickeringii var. pickeringii	Pickering's morning glory		R
	Symphyotrichum georgianum	Georgia aster	C	R
	Symphyotrichum novae-angliae	New England aster		R
	Tephrosia chrysophylla	sprawling goats rue		
	Thalia dealbata	powdery alligator-flag		T
	Thaspium chapmanii	creamy meadow-parsnip		T
	Tillandsia bartramii	Bartram's airplant	C	T
	Tillandsia recurvata	ball-moss		
	Tillandsia setacea	needle-leaf airplant		

FEDERAL STATUS LE—listed as endangered LT—listed as threatened C—candidate for listing under the Endangered Species Act
GEORGIA STATUS E—endangered T—threatened R—rare U—unusual

GROUP	SCIENTIFIC NAME	COMMON NAME	FEDERAL STATUS	GEORGIA STATUS
Vascular Plants *(continued)*	*Tragia cordata*	heartleaf nettle vine		
	Trillium lancifolium	lanceleaf trillium		
	Trillium reliquum	relict trillium		
	Vicia minutiflora	pygmy-flower vetch	LE	E
	Vigna luteola	wild yellow cowpea	LE	E
	Vitis palmata	riverbank grape	LE	E
	Waldsteinia lobata	barren strawberry	LE	E
	Zamia integrifolia	Florida coontie	LE	E

Acknowledgments

We three authors are grateful to the University of Georgia Press for its enthusiasm for this project, especially Nicole Mitchell and Laura Sutton, who were on board from the start. Thanks also to Judy Purdy, John McLeod, and the entire wonderful staff at the Press for their energy, creativity, and dedication. Thanks to Molly Thompson for her insightful editing.

Thanks to Craig and Diana Barrow of the Wormsloe Foundation for helping to bring this book into being.

We thank Altamaha Riverkeeper, which was the vehicle through which these photographs came about. Jeannie Lewis and Brad Winn gave James his first impromptu lesson in photography, and a grant from the Georgia Department of Natural Resources Nongame Wildlife Program provided a good camera.

Many thanks go to Altamaha Riverkeeper's executive director, Deborah Sheppard, and to staff over the years—Bryce Baumgartner, Debbi Davis, Donna Davis, Ben Emanuel, Wendy Vasquez Galan, Gail Krueger, Jim Kulstad, Billie Jo Parker, Constance Riggins, Matthew Teti, and John Wilson. We thank lobbyist Neill Herring for his advocacy on behalf of Georgia's rivers. James is especially grateful to Constance Riggins, also a native of the watershed, who assisted him for many years and who is a fond and robust supporter of the river.

All of Altamaha Riverkeeper's members, supporters, and volunteers, past and present, deserve a resounding "thank you" for their labors on behalf of the basin. Board members over the years include Tracey Adams, Kenny Atwood, Bruce Berryhill, E. A. Cheek, Frankie Clark, Jeannine Cook, Irwin Corbitt, Richard Creswell, Robert DeWitt, Bill Duckworth, Kyla Hastie, Len Hauss, Neill Herring, Carolyn Hodges, Wright Gres, David Kyler, Christi Lambert, Marilyn Lanier, Richard Madray, John Pasto, Carl Poppell, Janisse Ray, Gordon Rogers, Jack Sandow, Ann Trapnell, Albert Way, Mary Ellen Wilson, and Whit Perrin Wright.

A percentage of the proceeds from this book will go to support the work of the Altamaha Riverkeeper.

Many, many folks provided opportunities for James to get in the field; they include: Wendell Berryhill, Joe and Jane Fulcher, Faye Hinson, Mitch King, Carmen and Terye Trevitt, and Walt and Becky Woods. Thanks to Rick Dove for inspiration and support.

SouthWings, an environmental flying service based in Asheville, North Carolina, took James soaring through the heavens on many occasions. From the cockpit during these flights, James snapped some outstanding images. The work of SouthWings is integral to protecting the southern environment, and our gratitude is immense.

Dorinda Dallmeyer would like to thank Byron J. Freeman, director of the Georgia Museum of Natural History; Merryl Alber and Daniela De Iorio of the University of Georgia School of Marine Programs; and Tom Patrick, Brad Winn, and Tim Keyes of the Georgia Department of Natural Resources for technical advice on her essay, the image captions, and the Altamaha species lists. Of course, any errors or omissions are the author's alone. Dorinda also extends her thanks to Steven W. Engerrand of the Georgia Archives and Charles B. Greifenstein of the American Philosophical Society for images appearing in her essay.

Photographer Nancy Marshall of McClellanville, South Carolina, took the lovely black-and-white photo of James in the old-growth cypress of the Murff Tract. We are glad she granted us permission to use it here.

Our deepest appreciation goes to our loving spouses, Sumiko Holland, David Dallmeyer, and Raven Waters, as well as to our beautiful children and grandchildren.

We thank each other for the pleasure of working together. More than anything, we are grateful, each of us, to have been born within the Altamaha rivershed and to have been able, during our lifetimes, to call it home.

Index